Recent Themes in the History of Science and Religion

Historians in Conversation:
Recent Themes in Understanding the Past
Series editor, Louis A. Ferleger

Recent Themes in

THE HISTORY OF SCIENCE AND RELIGION

Historians in Conversation

Edited by Donald A. Yerxa

The University of South Carolina Press

Published by the University of South Carolina Press
Columbia, South Carolina 29208

www.sc.edu/uscpress

Manufactured in the United States of America

18 17 16 15 14 13 12 11 10 09   10 9 8 7 6 5 4 3 2 1

Library of Congress Cataloging-in-Publication Data

Recent themes in the history of science and religion : historians in conversation / edited by Donald A. Yerxa.
p. cm. — (Historians in conversation)
Includes bibliographical references and index.
ISBN 978-1-57003-870-9 (pbk : alk. paper)
1. Religion and science—Historiography. I. Yerxa, Donald A., 1950–
BL245.R43 2009
201'.6509—dc22

2009022161

This book was printed on Glatfelter Natures, a recycled paper with 30 percent postconsumer waste content.

Contents

Series Editor's Preface

The Historical Society was founded in 1997 to create more venues for common conversations about the past. Consequently in the autumn of 2001, the Historical Society launched a new type of publication. The society's president, George Huppert, and I believed that there was an important niche for a publication that would make the work of the most prominent historians more accessible to nonspecialists and general readers. We recruited two historians who shared this vision, Joseph S. Lucas and Donald A. Yerxa, and asked them to transform *Historically Speaking* into a journal of historical ideas. Up to that point *Historically Speaking* had served as an in-house publication reporting on the society's activities and its members' professional accomplishments. Yerxa and Lucas quickly changed the layout and content of *Historically Speaking,* and soon many of the most prominent historians in the world began appearing in its pages—people such as Danielle Allen, Niall Ferguson, Daniel Walker Howe, Mary Lefkowitz, Pauline Maier, William McNeill, Geoffrey Parker, and Sanjay Subrahmanyam. *Historically Speaking*'s essays, forums, and interviews have drawn widespread attention. The *Chronicle of Higher Education*'s Magazine and Journal Reader section, for example, repeatedly highlights pieces appearing in *Historically Speaking.* And leading historians are loyal readers, praising *Historically Speaking* as a "must-read" journal, a "*New York Review of Books* for history," and "the most intellectually exciting publication in history that is currently available."

The Historical Society is pleased to partner with the University of South Carolina Press to publish a multivolume series *Historians in Conversation: Recent Themes in Understanding the Past.* Each thematic volume pulls key essays, forums, and interviews from *Historically Speaking* and makes them accessible for classroom use and for the general reader. The original selections from *Historically Speaking* are supplemented with an introductory essay by Yerxa and suggestions for further reading.

We welcome your interest in the Historical Society. You may find us on the Internet at https://www.bu.edu/historic/. You may also contact us at the Historical Society, 656 Beacon St., Mezzanine, Boston, Mass., 02215-2010, telephone 617-358-0260.

Louis A. Ferleger

Acknowledgments

This volume tackles some "big questions" at the interface of science, religion, and history. The conversations printed here are selected from a series of forums appearing in *Historically Speaking* from 2005 to 2008. The series was generously funded by a grant from the John Templeton Foundation. I wish to thank Charles Harper and Paul Wason at the Templeton Foundation for their support and encouragement.

I have been blessed to work with wonderful colleagues at *Historically Speaking* and the Historical Society. It is such a pleasure working with my fellow editors Joe Lucas and Randall Stephens. Both are gifted historians and first-rate editors. It has also been my privilege to work with the society's founding executive director, Lou Ferleger, who recently retired after ten years of dedicated service.

I am indebted to the superb scholars whose work appears in this volume. They include many of the leading historians of science who are interested in the engagement of science and religion. I have worked with a number of the contributors over the last decade on a variety of projects, and several are good friends. In particular, I thank William R. Shea, the Galileo Professor of History of Science at the University of Padua. I met Bill at Oxford University in 1999 when I was participating in an intensive three-summer program in science and religion. Bill, the world's leading Galileo scholar, was one of the featured distinguished speakers. Since then, we have collaborated on a number of writing and professional projects that include, most recently, helping to organize a summer school at the Istituto Veneto di Scienze, Lettere ed Arti. I cannot thank him enough for the many enriching professional opportunities and personal kindnesses he has extended to me.

Introduction

Big Questions and the Complex Engagements of Science and Religion in History

Donald A. Yerxa

Where did we come from? What are the meaning and purpose of life? Where are the cosmos and humanity heading? Asking these and other "big questions" is basic to humanity. Religion represents the most common and persistent way humankind has addressed such questions, albeit with many and often conflicting variations. Science has opened other avenues of inquiry using methodologies that feature critical examination of evidence while ruling out such things as divine wild cards and special revelations. Science and religion do not exhaust the ways humans have addressed the big questions. There are, of course, the arts. But these two ways of knowing have had enormous influence on the course of history and the shape of modern thinking. Consequently the nature of the interactions between science and religion over the last several centuries is of great interest to historians.

Alfred North Whitehead once claimed that the future of civilization depends upon the way the two most powerful forces of history, science and religion, settle into relationship with each other. Judging from the ways that many public intellectuals, journalists, and even some historians have treated the interaction between science and religion, our future may be quite dim. The perception that science and religion, especially Christianity, have been engaged in centuries of conflict remains robust, fueled by the ongoing controversy over evolution and repeated references to historical episodes like the Galileo affair, the Huxley-Wilberforce debate, and the Scopes trial.

For decades the John Templeton Foundation has funded research and promoted inquiry on the intersection of science and religion. More recently the foundation has used the label of *big questions* to describe its philanthropic endeavors. Science and religion now join such core themes as humility, human flourishing, progress, purpose, ultimate reality, and unconditional love under Templeton's big-questions umbrella.[1] In 2004 the Historical Society received a grant from the Templeton Foundation to publish a number of forums in the pages of *Historically Speaking* on the big questions, especially related to the historical aspects of the interface between science and religion. The present volume reproduces four of these and boasts an outstanding roster of participating scholars. The forums have the same structure. Each begins with a lead essay by a scholar of international reputation. Other prominent scholars respond followed by a concluding rejoinder by the lead essayist. It must be emphasized that although these forums were made possible by a grant from the Templeton Foundation, at no point did the foundation or this editor attempt to steer the conversation to some prescribed conclusion.

Part 1 explores the historical relationship of science and religion in general. Its starting point is the notion that gained currency during the Victorian era that science and religion, particularly Christianity, were implacable foes. Despite the amazing resiliency of this warfare or conflict master narrative—no doubt fueled in the United States by seemingly endless debates over evolution—a rich, historiographical literature has convincingly demonstrated its inadequacy. As William R. Shea, who holds the Galileo Chair at the University of Padua, notes in his lead essay, historians have exposed a number of myths (for example, Galileo was tortured; Thomas H. Huxley defeated Bishop Samuel Wilberforce in their famous Oxford debate in 1860). Unfortunately it is a literature that by and large has not penetrated into history textbooks, let alone the public consciousness.[2]

How do we account for the persistence of the science-and-religion-in-conflict trope in the light of scholarship that has all but demolished the notion of a fundamental antagonism between two monolithic forces of science and religion? To help sort through these matters, Shea assesses the historical relationship between science and Christianity. Two prominent American historians of science, Ronald L. Numbers and Edward J. Larson, respond to Shea, who concludes this section with a rejoinder.

In his response to Shea, Numbers criticizes the assumptions that there is a singular historical relationship between science and religion and that science and religion "represent universal and unchanging categories." Numbers calls attention to some crucial features of contemporary scholarship that are reflected in the all the forums in this volume: Context and place matter, contingency reigns, and historical complexity abounds. Most historians of science and religion are antiessentialists, convinced of the necessity of close examination of the particulars.

Interestingly, while historians have been highly successful at demolishing the old warfare metanarrative along with a string of corollary myths, they have been less successful at constructing alternative models that embrace the complexities of the many engagements, large and small, of science and religion. Part 2, which features John Hedley Brooke's case for historical complexity, touches on this. It is no surprise, as the distinguished historian of science Brooke notes, that in a time when historians are deeply suspicious of metanarratives, they have emphasized historical complexity. Nowhere is this more evident than in the history of science and religion, a subject to which Brooke has made substantial contributions. Leery of the warfare metaphor, historians of science and religion have adopted an antiessentialist approach, dubbed the "complexity thesis," functionally elevating it to near metanarrative status. Although the complexity thesis is often associated with Brooke's work, he himself rejects the term. Complexity is an inadequate conceptual replacement for the discredited conflict metanarrative. It is, as Brooke observes, not a thesis but a description of historical reality. Close examination of particular historical contexts does not reveal "some timeless inherent relationship" between science and religion. Some scholars are concerned that highlighting complexity tends to reinforce fragmentation and incoherence in historical inquiry. Brooke and Geoffrey Cantor caution in their Gifford Lectures that preoccupation with particular contexts might dissolve "the great issues that have been debated under the banner of 'science and religion' into the fragments of local history."[3] But Brooke maintains that paying attention to complexity does not eliminate the historical patterns needed to make coherent historical narratives; it just yields ones that are more intricate.

Behind the specific issues at stake for historians of science and religion loom very important questions that cut to the core of contemporary historical inquiry. Does complexity, with its insistence on the local and the particular, necessarily run counter to efforts to synthesize and look for patterns? Can historians function without resort to master narratives, even though they necessarily blur the details of historical contexts? The tension between an emphasis on complexity and the concern over fragmentation in the absence of satisfactory master narratives surfaces frequently in historical inquiry these days. Nowhere is this more evident than in current discussions of the Scientific Revolution. This is the focus of part 3, in which the leading historian of the rise of science, Peter Harrison, revisits British historian Herbert Butterfield's claim that science gives the modern West its distinctive character.

As Shea notes in his response, the notion of a scientific revolution running from Nicolaus Copernicus to Sir Isaac Newton only gained currency a relatively short time ago with the publication in 1949 of Butterfield's classic *The Origins of Modern Science.* Recent historiography, however, has not been kind to the concept of a coherent and momentous scientific revolution.[4] Indeed Harrison contends that lately the Scientific Revolution has undergone something of an identity crisis. To be sure, it survives in undergraduate textbooks and survey courses, but it is frequently dismissed in more specialized work. More than fifty years ago Butterfield declared in *Origins of Modern Science* that the Scientific Revolution of the sixteenth and seventeenth centuries was a major formative influence on Western modernity: "Since that revolution overturned the authority in science not only of the middle ages but of the ancient world—since it ended not only in the eclipse of scholastic philosophy but in the destruction of Aristotelian physics—it outshines everything since the rise of Christianity and reduces the Renaissance and Reformation to the rank of mere episodes, mere internal displacements, within the system of medieval Christendom" (vii). Commenting on Harrison are three veteran historians of science, Charles Gillispie, David Lindberg, and Shea.

In recent years a number of historians have claimed that there was no such thing as the Scientific Revolution. Steven Shapin, for example, begins his revisionist account of seventeenth-century science with the memorable line: "There was no such thing as the Scientific Revolution, and this is a book about it."[5] Though his overall argument was considerably less provocative than this opening line might suggest, Shapin denied that there was anything like "an *essence* of the Scientific Revolution" and emphasized "the heterogeneity, and even the contested status, of natural knowledge in the seventeenth century."[6] A decade later the editors of *The Cambridge History of Science* volume dedicated to the early modern period intentionally avoided the phrase "the Scientific Revolution" in its title. In their introductory essay Katharine Park and Lorraine Daston observe that "the cumulative force of the scholarship since the 1980s has been to insert skeptical question marks after every word of this ringing three-word phrase, including the definite article. . . . It is no longer clear that there was any coherent enterprise in the early modern period that can be identified with modern science, or that the transformations in question were as explosive and discontinuous as the analogy with political revolution implies, or that the transformations were unique in intellectual magnitude and cultural significance."[7]

Although they may want to refine or reconceptualize the traditional framework, none of the distinguished historians who participate in the third forum is prepared to "derevolutionize" early modern science.[8] Their views represent a reasoned set of responses to the notion that the Scientific Revolution is no longer a useful construct. They also reveal the importance and functionality of the search for coherence[9] in a historiographical climate that celebrates novelty, the particular, the local—in a word, complexity. As already mentioned, Brooke and others in the current volume make a convincing case for historical complexity. But H. Floris Cohen in an essay published elsewhere rightly laments that sometimes the novelty and complexity of the new scholarship on the Scientific Revolution have been accompanied with a skeptical resignation that there can be an underlying coherence to the various developments of seventeenth-century science/natural philosophy.[10] This is no small matter. If the quest for a coherent scientific revolution is deemed a fool's errand, what then of other historical frameworks like the Renaissance and the Enlightenment? Indeed can any framework withstand the onslaught of those who might be called "resigned complexifiers"? It is not, of course, inappropriate to question the coherence of a given framework or periodization scheme. And, as Brooke argues, it is wrong to link complexity with the kind of skeptical resignation that troubles Cohen. Brooke offers a vastly more constructive response to the recent scholarship: Use its findings to test, refine, and, if need be, replace existing frameworks with new ones that offer more coherent renderings of the past. The only question that remains, to my mind, is whether historians will be bold enough to advance new frameworks, even on a provisional basis, in the current intellectual climate of the academy.

This brings us back to the matter of big questions. Part 4 concludes the current volume with a discussion of one of these "*really* big questions": Is there progress in history? Here our guide is Bruce Mazlish. One of the foremost experts in the history of the social sciences, Mazlish, who with Leo Marx edited *Progress: Fact or Illusion?* (1996), explores the issues of progress and teleology. A panel of distinguished scholars responds, followed by Mazlish's rejoinder. The idea of progress dates from antiquity and ranks, according to Robert Nisbet, among the most important ideas produced by Western civilization.[11] Similarly, one of the leading philosophers and intellectual historians of our time, Leszek Kolakowski, considers progress and its relation to morality to be one of humanity's most intriguing and significant questions.[12] If for whatever reason historians cease to traffic in explanatory schemes for big questions like those that lie behind the concept of the Scientific Revolution, they will have reneged on an important social responsibility—to make sense of the past—and risk forfeiting their claim for a valued place in the academy. That said, another category of big questions regarding the past—*really* big questions—engages us as humans, but most historians consider this group beyond their disciplinary warrant. Historians prefer to address manageable topics suited to their method and practice—a commonsensical approach of interrogating reliable evidence and constructing narratives that make the best inference to explanation. They gravitate to what John Demos has called "specified particulars—this time, that place, these events."[13]

Notions of progress and teleology have been all but banished from contemporary historical interpretation. But those historians who examine the past in large chunks seemingly cannot avoid the obvious: The past reveals a general trajectory of increasing social and economic complexity. Discussions of evolutionary biology encounter the same tensions between contingency and directionality. Can historians legitimately incorporate notions of complexity and directionality without taking on the unwanted baggage of progress?

Contemporary historians, however, avoid the subject of progress.[14] The reason for this is no mystery. Progress smacks of teleology—historians prefer the term *Whiggishness*—that is anathema to most historians. Though not a historian, political philosopher John Gray captures well the deep reservations historians have about invoking progress. Echoing the great historian of science George Sarton, Gray proclaims that only in the area of science—which, presumably, includes medicine and technology—has there been genuine progress. In ethics or politics, he claims, progress is an illusion. Of course there have been "gains" over time, but these are not cumulative. Improvement, Gray argues, can never be cumulative: "What has been gained can also be lost, and over time surely will be."[15] Contingency and human agency—definitely not teleology—drive the engines of historical inquiry today.

Historians often recite a sad litany of the horrific violence of the past century to mock naïve notions of progress. How can one argue for progress after the carnage of two world wars, the Holocaust, and other depressing episodes of ethnic or ideological cleansing? Beyond all this what (or whose) yardstick will historians use to measure progress? Case closed. Well, maybe not. Some leading economists, for example, Nobel laureates Robert Fogel and Robert E. Lucas Jr. remind historians they must consider the substantial empirical evidence of such things as population growth, infant mortality, body heights and weights, income per person, and other indicators that reveal statistically dramatic improvement in human well being, especially in the last century.[16]

A few historians have broken ranks. David Hackett Fischer, for example, boldly concludes that the growth of liberty and freedom is "the central theme of American history." His examination of the evidence convinces him that over the past four centuries, "every American generation without exception has become more free and has enlarged the meaning of liberty and freedom in one way or another." Making the case for directionality in history, as Fischer does, is not the same thing as arguing for determinism and teleology. But he is clearly embracing a version of progress compatible with contingency.[17] And Wilfred M. McClay asserts that belief in progress is so "thoroughly inscribed in our cultural makeup," it may well be inescapable. What we must do, he wisely contends, is to stop distancing ourselves from the notion with our "sneer quote" mentality and "find better ways of talking about it and thinking about it, ways of chastening it, restraining it, and protecting it against its excesses"[18] The concluding forum in this volume represents a constructive step in the direction McClay advocates.

Herbert Butterfield himself once observed that the big questions do not go away merely by ruling them off limits.[19] The forums that are printed in this volume demonstrate the value of examing the big questions from historical perspectives even though academic and disciplinary fashions seem to favor avoidance.

Notes

1. For further information, see the John Templeton Foundation Web site, http://www.templeton.org/funding_areas/core_themes/. Keith Ward, the distinguished Oxford Regius Professor of Divinity Emeritus, has written a book published in 2008 by the Templeton Foundation Press on *The Big Questions in Science and Religion.* Ward's list of big questions includes: How did the universe begin? How will the universe end? Is science the only sure path to truth? Has science made belief in God obsolete?

2. See Jon Roberts, "'The Idea That Wouldn't Die': The Warfare between Science and Christianity," *Historically Speaking* 4 (February 2003): 21–24.

3. John Brooke and Geoffrey Cantol, *Reconstructing Nature: The Engagement of Science and Religion* (Edinburgh: T&T Clark, 1998, 25.

4. The following comments on the Scientific Revolution are based on Donald A. Yerxa, "Historical Coherence, Complexity, and the Scientific Revolution," *European Review* 14 (October 2007): 439–44.

5. Steven Shapin, *The Scientific Revolution* (Chicago: University of Chicago Press, 1996), 1.

6. Ibid., 161–62.

7. Katharine Park and Lorraine Daston, "Introduction. The Age of the New," *The Cambridge History of Science: Volume 3, Early Modern Science,* ed. Park and Daston (Cambridge: Cambridge University Press, 2006), 2–3.

8. See Peter Anstey's review essay "Derevolutionizing Early Modern Science," *Metascience* 17 (2008): 389–96.

9. For a superb discussion of the notion of coherence in historical study, see Allan Megill, "Coherence and Incoherence in Historical Studies. From the Annales School to the New Cultural History," *New Literary History* 35 (2004): 207–31.

10. H. Floris Cohen, "Reconceptualizing the Scientific Revolution," *European Review* 14 (October 2007): 491–502.

11. Robert Nisbet, *History of the Idea of Progress* (New York: Basic Books, 1980), 4.

12. Leszek Kolakowski, *Why Is There Something Rather Than Nothing?* (New York: Basic Books, 2007), esp. 172–73.

13. John Demos, "Response to Adam Hochschild," *Historically Speaking* 9 (March/April 2008): 7.

14. There have been a few scholarly explorations of the idea of progress in recent decades. See Nisbet, *History of the Idea of Progress;* and Christopher Lasch, *The True and Only Heaven: Progress and Its Critics* (New York: Norton, 1991).

15. John Gray, *Heresies: Against Progress and Other Illusions* (London: Granta, 2004), 2–4. For a challenging discussion of the notion of scientific progress and the case for noncumulative, "problem-solving scientific progressiveness," see Larry Laudan, *Progress and Its Problems: Towards a Theory of Scientific Growth* (Berkeley: University of California Press, 1977), esp. 1–8, 147–50, and 223–25.

16. For a perspective that acknowledges these indexes of progress but interprets intensified periods of transformation in the light of the "spiritual struggles" they generate, see Robert William Fogel, *The Fourth Great Awakening and the Future of Egalitarianism* (Chicago: University of Chicago Press, 2000), esp. 236–42.

17. David Hackett Fischer, *Liberty and Freedom: A Visual History of America's Founding Ideas* (New York: Oxford University Press, 2005), 721–22. See also Donald A. Yerxa, "David Hackett Fischer's *Liberty and Freedom* in Historiographical Perspective," *Historically Speaking* 7 (September/October 2005): 16–17.

18. Wilfred M. McClay, "Revisiting the Idea of Progress in History," *Historically Speaking* 9 (September/October 2007): 12.

19. Herbert Butterfield, *Man on His Past* (Cambridge: Cambridge University Press, 1955), 141. This volume comprises Butterfield's Wiles Lectures at Queen's University, Belfast, Ireland, 1954.

PART 1

Science and Religion in Historical Perspective

Assessing the Relations between Science and Religion

William R. Shea

We inhabit a present-tense culture. Only when we grasp that some historical event has been understood in a variety of ways do we come to realize that our current viewpoint is not necessarily anchored and cemented in hard facts. In the seventeenth century Protestant reformers used the Roman Inquisition condemnation of Galileo for teaching that the Earth moves to deal a blow to the claims of the papacy. One hundred years later the secular Enlightenment turned the Galileo affair into a stick to trounce Christianity in general, and by the nineteenth century it had become a battleaxe to shatter any kind of religion. Those who shared this antireligious view believed that progress was only possible if the human mind was liberated from the trammels of religious creeds and made to rely exclusively on science, the embodiment of rationality.

No characterization of the relationship between science and religion has proved more seductive and tenacious than that of conflict. Indeed the two classic works on the subject are entitled *History of the Conflict between Religion and Science* (1824) and *A History of the Warfare of Science with Theology in Christendom* (1896). The former, which passed through twenty editions and was translated into nine languages, was written by the chemist-historian John William Draper, who saw Christianity, and especially Roman Catholicism, as the archenemy of science. "The history of Science," he wrote, "is not a mere record of isolated discoveries; it is a narrative of the conflict of two contending powers, the expansive force of the human intellect on one side, and the compression arising from traditionary faith and human interests on the other."[1] The author of the second work, Andrew Dickson White, was

From *Historically Speaking* 7 (November/December 2005)

equally convinced that war was inevitable. As the first president of Cornell University, White was dedicated to creating a center of higher knowledge free from the constraints of religious creeds. In an age when the American establishment was still largely Protestant, this generated opposition, and White came to see himself as another Galileo battling the arrayed forces of obscurantism. In his view the Yankee divines who objected to his plans behaved like the Italian clerics who had persecuted Galileo. White was confident that he would overcome, and, more important, he nurtured the hope that this would be during his lifetime and not after his death like Galileo.

Both Draper and White acknowledged that the historian must enter into the minds that he studies and that this requires an appreciation of the ideas, ambitions, and prejudices of the past. What they failed to grasp is that this can only be achieved if the historian is critically aware of his own ideas, ambitions, and prejudices. Self-knowledge is difficult at all times, but Draper and White had swallowed whole a view that made it impossible for them to exercise self-criticism. They believed that the scientific method ushered in by the Scientific Revolution provided them with a way of understanding not only nature but history. This view thrived well into the first half of the twentieth century and received its canonical formulation (and geometrical garb) in George Sarton's *Study of the History of Science,* published in 1936:

Definition: Science is systematized positive knowledge or what has been taken as such at different ages and in different places.
Theorem: The acquisition and systematization of positive knowledge are the only human activities which are truly cumulative and progressive.
Corollary: The history of Science is the only history which can illustrate the progress of mankind. In fact, progress has no definite and unquestionable meaning in fields other than the field of science.[2]

Sociologists of science enthusiastically embraced the notion that science possesses a unique set of progressive norms. In a characteristic utterance, Robert K. Merton listed these norms as "universalism, communism, disinterestedness, and organized scepticism,"[3] as if there was nothing incongruous in lumping together intellectual, moral, and political qualities. This consensus combined the hallmark of dogma with the messianic promise of a good life for all, and it was slow to be challenged because its adepts, when criticized, were able to shift their ground from the empirical side of scientific theories to the logical side or vice versa. In the end they were routed on both fronts. On the empirical side it was gradually realized that "hard facts" represent more an ideal than a reality. The unfolding of science could be written as the history of a debating society discussing such topics as: Is the Earth or the Sun at rest? Do bodies fall because they are heavy or because they are attracted? Is the present shape of the Earth the result of gradual or sudden change? Was the agent of change water or fire? Do we need one or two fluids to explain electricity? Is combustion due to the release of phlogiston or the absorption of oxygen? Are acquired characteristics transmitted to offspring or are all mutations random? Is light a corpuscular or an undulatory phenomenon? Is quantum physics really indeterministic? All these debates involved prominent scientists who offered genuinely different theories while ostensibly accounting for the same empirical data. Equally important is the fact that these discussions often went on for decades. For instance, distinguished scientists such as Johann Bernoulli and Pierre-Louis Moreau de Maupertuis were still weighing the pros and cons of Newtonian gravitation fifty years after Newton's *Principia Mathematica* had appeared. Only a short-term, short-lived view creates the impression of consensus, for scientists like all men,

> are here as on a darkling plain
> Swept with confused alarm of struggle and flight,
> here ignorant armies clash by night.[4]

Scientific methodology fared hardly better on the logical side, where controversy is not only rampant but inevitable because the rules of evidence do not and cannot pick out one theory unambiguously to the exclusion of all others. From the 1950s onward, the argument took two forms: One is associated with the names of Pierre Duhem and Willard Van Orman Quine and aims to show that no theory can be logically proved or refuted by any body of evidence; the other is displayed in the writings of Nelson Goodman, who criticizes the rules of scientific evidence as radically open to different and even inconsistent interpretations. A more radical version of this problem was made fashionable by Paul Feyerabend, for whom rival theories, if they are more than trivially different accounts of the same hypothesis, each belong to different conceptual universes.[5]

The realization that scientific data are underdetermined and that theories are partly incommensurable contributed to the end of blind faith in the benign influence of science and technology. This at a time when everyone suddenly became morally accountable: the engineer, who was told that it is not enough to design an efficient plant as he had been trained to do but that he had to gauge its impact on the environment; the administrator, who was asked to consider the long-range consequences of computerization on personnel; the manager, who had to assess the social implications of automation in the factory; the doctor, who was faced with new ethical problems posed by a technology that enabled him/her to initiate or prolong life almost at will; the parents, who had to decide what additives are acceptable in the food they gave their children; and the banker, who was expected to respect privacy and safeguard the computerized network from becoming a vehicle for fraud. The waning of science as a charismatic profession capable of training men and women to look at nature in an objective and self-detached way led to a surprising development. Science had thought that it was seeing religion out the back door. But while it was distracted by questions that lay beyond its ability to resolve, religion and other forms of ethical discourse quietly walked back in through the front door.

The notion of goodness has been discussed since the Greeks invented philosophy over two thousand years ago, but our vision of the world never was and never will be completely shaped by a debating society, however sharp, witty, or wise. Human beings need the experience and the support of systems that can cope with their emotional as well as their intellectual needs. The great religions provide insights that address these emotional needs.

In this new climate historians are at liberty to reexamine the Galileo affair. No longer under compulsion to hail it as a glorious episode in the battle of scientific light against religious darkness, they have taken a closer look at the historical context and the persons involved. Religion, in the twenty-first century, is asked to show its credentials, and there is little agreement on what counts as genuine, as distinguished from putative, insight. The reverse was the case in the seventeenth century when virtually everyone took it for granted that the Bible was literally true unless there was a compelling reason for considering some passages as mere figures of speech. Scripture mentions that the Sun rises and sets, and since this agrees with everyday observation, it was reasonable to believe one's senses. Galileo was convinced that he could prove that the Earth, not the Sun, is in motion. His main argument was that tides could not occur in the oceans unless the Earth both rotated on its axis and revolved around the Sun. It was an ingenious argument. Unfortunately it was also completely wrong. Galileo trumpeted his spurious demonstration, and Pope Urban VIII clung just as passionately to the traditional view that the Bible was to be considered literally true unless shown otherwise.

Someone who tried to be faithful to Scripture while leaving room for an eventual reinterpretation of certain passages in the light of fresh scientific evidence was Cardinal Robert Bellarmine (1542–1621). When a friar named Paolo Antonio Foscarini sent him a pamphlet in which he argued that the motion of the Earth was not at variance with the teaching of the Bible, the

Catholic cardinal wrote back to express a view that was eminently sensible in the light of current beliefs. Bellarmine shared the same concerns as the Protestant theologians with whom he carried on what he called "an honest discussion" (although we might be tempted to describe some of it as "theological bickering"), and he was at one with them in stressing that Christianity is a revealed and, hence, a historical religion. The acts of God occurred at a given time and were recorded in a series of books that have a peculiar kind of veracity, which is conveyed by the word *inspired.* This meant, for Bellarmine and the generation prior to Galileo's telescopic discoveries, that apparently straightforward statements in the Bible are to be taken as such unless proved to have a different or a looser meaning. Here is how Bellarmine put it: "Someone who denied that Abraham had two sons and Jacob twelve would be just as much a heretic as someone who denied the virgin birth of Christ, for both are declared by the Holy Ghost through the mouths of the prophets and the apostles."[6] When Galileo read this, he commented, "It is much more a matter of faith to believe that Abraham had sons than that the Earth moves. . . . For since there have always been men who have had two sons, or four, or six, or none, . . .there would be no reason for the Holy Spirit to affirm in such matters anything contrary to truth. . . . But this is not so with the mobility of the Earth, this being a proposition far beyond the comprehension of the common people."[7]

We can appreciate Galileo's point today, but was it convincing in the seventeenth century? Bellarmine, however mistakenly, believed that the Old Testament book of Ecclesiastes had been written by Solomon, "who not only spoke by divine inspiration, but was also a man wise above all others, and learned in the human sciences and in the knowledge of all created things. This wisdom he had from God."[8] Nowadays we take our cue from astrophysicists; in the seventeenth century Bellarmine took it from Solomon. This is why when Bellarmine read the verse "The Sun rises and the Sun sets, and hurries back to where it rises" (Eccles. 1:5), he wanted to think twice before denying that it is literally true. He was willing, however, to acknowledge that if a genuine proof that the Earth moves were found, then we would have to reinterpret passages that appear to state that it is at rest. The rub is that neither Foscarini nor Galileo had such a proof. The argument from the tides did not hold water, as Bellarmine was told by the contemporary scientists whom he consulted.

Copernicus's book on the motion of the Earth had appeared way back in 1543, and the theory had been found mathematically interesting but physically out of step. There was a remote chance that it might turn out to be true in the long run, but until this was demonstrated, it was unwise to jettison the traditional reading of biblical passages that referred to the Sun as rising and setting, a way of speaking that remains deeply imbedded in our language. We go on talking about sunrise and sunset although we know that the Sun is at rest.

Religion's credentials may not be as appalling as Draper and White made them out to be, but our culture is so permeated by their assumptions that it is common to hear it stated that the Galileo affair was only one of many such incidents. Historians, in recent years, have exposed a number of such myths: the Church prohibited autopsies and dissections in the Middle Ages and the Renaissance; medieval Islamic culture was inhospitable to natural philosophy; Giordano Bruno was burned alive because of his acceptance of Copernicanism and his advocacy of an infinite universe; Galileo was tortured; and Darwin's "bulldog," Thomas Huxley, soundly defeated Bishop Samuel Wilberforce at a notorious meeting in Oxford in 1860.[9] By reining in their judgmental impulse, historians have elucidated the past. This is, of course, their job. But I would like to suggest that they have not only provided an insight into the way earlier generations developed different forms of knowledge but have also disseminated a methodological outlook that is sorely needed at the present time. The return of religious fundamentalism in our own day poses a threat because it rests on a scenario of bad guys and good guys. Historians may no longer be avengers, but they should be allowed a modest plea in favor of understanding and tolerance.

Notes

1. John William Draper, *History of the Conflict between Religion and Science* (London: King, 1875), vi.

2. George Sarton, *The Study of the History of Science* (Cambridge, Mass.: Harvard University Press, 1936), 5. The book was reprinted as late as 1957.

3. Robert K. Merton, *Sociology of Science* (Chicago: University of Chicago Press, 1973), 268–69.

4. From the last three verses of Matthew Arnold's poem "Dover Beach."

5. Quine's version of Duhem's thesis is that "our statements about the external world face the tribunal of sense experience not individually but as a corporate body." Willard Van Orman Quine, "Two Dogmas of Empiricism," in *From a Logical Point of View* (Cambridge, Mass.: Harvard University Press, 1959), 41–43. The consequences he draws from this thesis are the following: (a) it is misleading to speak of the "empirical content" of an individual statement; (b) any statement can be retained as true provided that sufficiently drastic adjustments are made elsewhere in the system; (c) there is no sharp boundary between synthetic statements whose truth is contingent upon empirical evidence and analytic statements whose truth is independent

of empirical evidence. For Nelson Goodman's views, see mainly *Fact, Fiction, and Forecast*, 4th ed. (Cambridge, Mass.: Harvard University Press, 1983). Paul Feyerabend's popular *Against Method* first appeared in 1975.

6. Cardinal Robert Bellarmine to Paolo Antonio Foscarini, April 12, 1615, *Discoveries and Opinions of Galileo,* trans. and ed. Stillman Drake (New York: Doubleday, 1957), 163.

7. Ibid., 168.

8. Ibid., 164.

9. Two American historians, David C. Lindberg and Ronald L. Numbers, have played a significant role in encouraging scholars to contextualize and reassess the relations between science and faith. The shift from cold war to frank dialogue can be seen by comparing the essays that they edited and published in 1986, *God and Nature: Historical Essays on the Encounter between Christianity and Science* (Berkeley: University of California Press), with the new collection that they edited in 2003, *When Science and Christianity Meet* (Chicago: University of Chicago Press).

Response to William Shea

Edward J. Larson

William Shea explains how the Galileo affair has served over the past 350 years as the principal evidence supporting what he describes as the "seductive and tenacious" view that conflict or warfare has characterized the relationship between science and religion. I certainly don't know as much about Galileo as Bill Shea does, and I readily accept the historical significance that he ascribes to the Roman Catholic Church's persecution of the Italian scientist. He concludes his essay on the hopeful note—hopeful that is for us pacifists and conflict avoiders—that recent developments in the sociology, philosophy, and history of science, coupled with postmodern reassessments of religion, point toward the demise of the warfare thesis and its replacement by a more contextualized view. "Historians may no longer be avengers," Shea concludes, "but they should be allowed a modest plea in favor of understanding and tolerance."

I'm all for understanding and tolerance. Indeed I rank them high among the spiritual and secular virtues. But I'm not so sure that very much has changed in how historians and scientists view either the Galileo affair in particular or the relationship between science and religion in general.

Like Shea, I'm not talking about an actual war between science and religion as if mental constructs can fight like countries do, marshalling armies and taking territory. We're talking about how people perceive the relationship between science and religion or, perhaps, what goes on in an individual's mind when confronted with the claims of science and religion. Warfare is a metaphor, of course, but the recognition that its application to the relationship between science and religion is both seductive and tenacious suggests that people respond to it and feel that it has validity.

From *Historically Speaking* 7 (November/December 2005)

Metaphors are thrown around all the time, but they only survive if they resonate with people. For example some American sports promoter once referred to the college basketball tournament as "March Madness," and the metaphor stuck. The tournament does not actually drive basketball fan "mad" in the clinical sense of that term, but the metaphor seemed apt to those who follow the sport. Once the phrase took hold, it helped to feed the "madness." Similarly, some nineteenth-century partisan spoke of "warfare" or "conflict" between science and religion, and that characterization seemed apt to enough people that it stuck and thereafter probably reinforced the sense of discord. So far as I can tell, the warfare metaphor remains as alive today as it was in the past.

Shea opens his essay by summarizing how the Galileo affair informed the science and religion debate over three centuries. In the seventeenth century, he says, Protestants used it to club Catholicism. Enlightenment secularists turned it on all Christianity during the eighteenth century. And in the nineteenth century it applied against religion generally. As evidence Shea cites two histories of the "conflict" written during the nineteenth century, one by John William Draper and the other by Andrew Dickson White.

I presume that Shea chose the books by Draper and White as his examples because they were written as works of history, and his essay is about historiography. It should be noted, however, that Draper and White were not primarily historians. Draper was a chemist, and White had become a college president. Other nineteenth-century histories presented religion as fostering science, with some Protestant historians claiming that the Reformation jump-started science and some Catholic historians praising the Roman Church's support for science. Thus, as history, the works of Draper and White were not necessarily representative of their time. Further, they did not so much attack religion generally (which Shea says was characteristic of the nineteenth-century depiction of the Galileo affair) as Catholicism in particular. This was especially true of Draper's book. The view that science was a war with all religion did not dominate nineteenth-century historical thought.

It should be added that the various nineteenth-century interpretations of the relationship between science and religion—from warfare to collaboration—did not solely appear in works of history. Scientists also expressed thes views, with a similar diversity of viewpoints as those expressed by historians. The warfare thesis appears in the writings of such prominent nineteenth-century scientists as Charles Lyell, Charles Darwin, Thomas H. Huxley, and Francis Galton. These British scholars typically focused their fury on Catholicism (often invoking the Galileo affair much as Shea depicts seventeenth-century Protestant reformers using it), but other religions could suffer their assaults as well. Huxley made this explicit in his published essays *Science and Christian Tradition:* "From the earliest times of which we have any knowledge, Naturalism [or science] and Supernaturalism [or religion] have consciously, or unconsciously, competed and struggled with one another." Protestants joined with secularists in using science to debunk Catholicism, Huxley notes, but "their alliance was bound to be of short duration, and, sooner or later, to be replaced by internecine warfare."[1] Here is the warfare thesis expressed by a combatant in the fray—and so it should carry considerable evidentiary weight.

At the same time, however, other scientists were expressing very different views of the relationship between science and religion. Think of Lord Kelvin, James Clark Maxwell, or Michael Faraday, for instance. These British physicists carried similar stature as Lyell, Darwin, and Huxley in the world of nineteenth-century science, yet none of them declared war on religion. Quite to the contrary, they saw science and religion as complementary. In America, the eminent nineteenth-century botanist Asa Gray lectured on the compatibility of Darwinian biology and biblical Christianity.

In short, although the warfare thesis was alive and well in nineteenth-century thought, it did not go unchallenged. Historians did not uniformly preach it and scientists did not uniformly confess it.

A similar diversity of opinion on the relationship of science and religion existed during the twentieth century and continues into the twenty-first century. At least in the United States, the period has witnessed a continuing divide between science and religion, with both flourishing in their separate spheres. Housed in ever-expanding research universities and fueled by unprecedented amounts of public funding, American science has assumed global leadership in virtually every scientific discipline. The technological payoff has transformed American industry, agriculture, and warfare. At the same time, surveys have found that a greater percentage of Americans regularly attend religious services and profess belief in God than the people of any other scientifically advanced nations—with no apparent decline over time. Yet surveys also suggest that these percentages drop off precipitously for American scientists, particularly at the higher echelons of the profession. This provides a context in which some conservative Christians can denounce objectionable scientific theories as the work of "atheistic scientists."

Perhaps the most significant development in the relationship of science and American religion over the past two centuries has been the disengagement of mainline Protestantism from the science-religion dialogue. In the wake of William Paley's popular works of natural theology, mainline Anglo-American Protestants regularly invoked science in support of their religious beliefs during the nineteenth century and sought to reconcile science with religion. In marked contrast, such preeminent twentieth-century Protestant theologians as Karl Barth, Paul Tillich, and Reinhold Niebuhr virtually ignored science in their theological writings. Mainline Protestants joined most Catholics in largely reserving their comments about science to ethical issues raised by technological applications of science (such as biotechnology or nuclear weapons) and to making the general observation that modern scientific theories (such as the Big Bang and quantum indeterminacy) leave room for God.

During the twentieth century, however, evangelical, fundamentalist, and Pentecostal churches have displaced mainline ones as the center of gravity within American Protestantism. Many in these churches feel that their beliefs are under siege from science—particularly from Darwinism but also from dominant theories in geology, cosmology, and psychology—and some of them militantly lash out against these threatening ideas. Lay Christians periodically over the past century have stirred mass movements against the theory of organic evolution. Presbyterian politician William Jennings Bryan did so in the 1920s, resulting in legal limits on the teaching of evolution in some public schools and the 1925 trial of high-school teacher John Scopes for violating one such law in Tennessee. Beginning in the 1960s, Baptist engineering professor Henry Morris helped to revive a literal reading among conservative Protestants of the Genesis account of creation, prompting widespread demands for teaching so-called creation science alongside Darwinism in biology classes. In the 1990s Presbyterian law professor Phillip Johnson rekindled interest among conservative Christians for pre-Darwinian concepts of intelligent design in nature, leading to demands that public schools modify the science curriculum to incorporate their concerns.

Each of these episodes has breathed new life into the warfare thesis and evoked comparisons to Galileo's persecution. Scopes's defenders frequently compared the young Tennessee school teacher to Galileo, for example, with his defense counsel at one point in the trial declaring that "every scientific discovery or new invention has been met by the opposition of people like those behind this prosecution who have pretended that man's inventive genius was contrary to Christianity." The lead prosecutor responded, "They say it is a battle between religion and science, and in the name of God, I stand with religion."[2] Similarly, both Morris and Johnson have presented Darwinian science as at war with Christianity, and some scientists have responded in kind. In *The Blind Watchmaker*, for example, British biologist Richard Dawkins takes aim at "redneck" creationists and "their disturbingly successful fight to subvert American education and textbook publishing."[3]

The warfare model still survives among historians as well. Shea rightly points out that some historians, encouraged by David C. Lindberg and Ronald L. Numbers, have begun reassessing the historical relations between science and religion, shifting from "cold war to frank dialogue." But other historians still speak in terms of a conflict between science and religion. Taking their cue from Morris and Johnson, for example, religious opponents of Darwinian naturalism interpret modern history in terms of a clash between Christian values and evolutionary ethics. Much of this is done by nonhistorians, such as Morris and Johnson, who themselves who blame all manner of modern "evil" from communism to total war on Darwinism, but their followers include trained historians, such as California State University history professor Richard Weikart, whose 2002 book, *From Darwin to Hitler,* roots Nazi barbarism in scientific naturalism and Hitler's repudiation of traditional Christian values. On the other side, even such a mainstream history of science text as *Science and Technology in World History* by James E. McClellan III and Harold Dorn (which I use in some of my introductory courses) perpetuates the warfare thesis, beginning with the comment that Thales, the first Greek natural philosopher, "sets the natural world off somehow separate from the divine" and continuing through vivid depictions of the Galileo affair and modern Christian opposition to Darwinism. For McClellan and Dorn, the conflict is between science and any authority that would "inhibit scientific development," with Christianity singled out for opposing theories of a moving Earth and evolving species.[4]

Given the shrill tone of the ongoing American controversy over creation and evolution, and the rising influence of both secular scientism and biblical religion in modern America, I cannot wholly dismiss the warfare metaphor as an antiquated relic. To partisans on both sides of this and other controversies, conflict remains an inevitable way to interpret the relationship between science and religion.

Notes

1. Thomas H. Huxley, *Science and Christian Tradition: Essays* (New York: Appleton, 1894), 5, 14.

2. *World's Most Famous Court Trial* (Dayton, Tenn.: Bryan College, 1990), 115, 183, 197. This is a published version of the Scopes's trial transcript.

3. Richard Dawkins, *The Blind Watchmaker* (Harlow, Essex, U.K.: Longman, 1986), 241.

4. James E. McClellan III and Harold Dorn, *Science and Technology in World History: An Introduction* (Baltimore: Johns Hopkins University Press, 1999), 61, 234, 329.

Comments on William Shea

Ronald L. Numbers

Perhaps surprisingly, given William R. Shea's area of expertise (early modern Europe) and his religious orientation (Roman Catholic), I find myself (a religiously agnostic historian of modern America), in almost total agreement. I especially appreciate what he and several other historians of science and religion have done to correct the notorious Galileo story. Thanks to them, we no longer see the Galileo affair as a paradigm of the inevitable conflict between science and theology but as a complex human interaction between the abrasive, egocentric, and fallible Galileo (who for years practiced and taught astrology) and some equally flawed (but no less intelligent) churchmen. Galileo no doubt suffered psychologically from his encounter with church authorities, but, contrary to nearly universal belief, he was neither tortured nor imprisoned.[1] In fact, no natural philosopher or natural historian, to my knowledge, ever lost his life because of his scientific views, though the Italian Inquisition did incinerate the sixteenth-century Copernican Giordano Bruno for his heretical *theological* notions.

In the nineteenth century John William Draper and Andrew Dickson White administered a severe historical beating to the Vatican, whose antipathy toward science had, in the words of Draper, left its hands "steeped in blood." The Roman church did not always support interrogating nature, but its record was far from the dismal indictment of Draper and White. In a recent history of solar observatories in cathedrals, the distinguished Berkeley historian John Heilbron concludes that "the Roman Catholic church gave more financial and social support to the study of astronomy for over six centuries, from the recovery of ancient learning during the late Middle Ages into the Enlightenment, than any other, and, probably, all other, institutions."[2] Heilbron's assertion, though counterintuitive to many readers, rests on sound

From *Historically Speaking* 7 (November/December 2005)

historical evidence. It provides no support, however, for the generalization, favored by some Christian apologists, that science could have developed only in a Christian culture. Such a claim misrepresents the scientific achievements of ancient Greeks and medieval Muslims and exhibits a thorough misunderstanding of the history of early Western science.

Although in broad agreement with Shea, I am somewhat uncomfortable with the editorially assigned task of discussing "mistaken historical interpretations of the relationship between science and religion." This phrasing, though common (I occasionally use it myself), unfortunately incorporates two of the problems that historians of the subject continue to face: (1) the expectation that we should be capturing a singular historical relationship—*the* relationship—between science and religion and (2) the assumption that *science* and *religion* represent universal and unchanging categories. The scholarship of the past quarter century has repeatedly revealed a complex entanglement of scientific and religious factors, which resists reduction to a single metaphor.[3] Although the term *science and religion* can sometimes serve as a convenient shorthand, it tends to reify both elements in ways that distort historical understanding.

Whose science are we talking about: the science of elites or of the people? Does the term embrace the so-called social sciences as well as the physical sciences? Does it include Christian Science and creation science as well as astronomy and zoology? George Sarton to the contrary notwithstanding, science does not float above historical contingency.

Until recently even historians of science paid scant attention to the changing and diverse meanings of *science,* which did not signify the study of nature until about two centuries ago. The shift from *natural philosophy* (and natural history) to *science* is of critical importance to the history of science and religion because natural philosophy, unlike science, did not rule out theological considerations. As Harvard University's Hollis professor of mathematics and natural philosophy John Winthrop informed the eighteenth-century readers of the *Boston Gazette,* "The consideration of a DEITY is not peculiar to *Divinity,* but belongs also to *natural Philosophy.* And indeed the main business of natural Philosophy is, to trace the chain of natural causes from one link to another, till we come to the FIRST CAUSE; who, in Philosophy, is considered as presiding over, and continually actuating, this whole chain and every link of it; and accordingly, I have ever been careful to give my discourses this turn."[4]

The very idea of writing about "science and religion" did not occur until about the 1820s, when Thomas Dick published *The Christian Philosopher; or,*

The Connection of Science and Philosophy with Religion (1823), the first book I've found that (almost) incorporates the phrase. Within a short time the term was appearing everywhere, in magazines, sermons, and the titles of professorships. Although the meaning of *science* remained ambiguous well into the nineteenth century—*Webster's Dictionary* in 1806 still defined it as "knowledge, deep learning, skill, art"—the word increasingly came to be associated with the natural sciences, both physical and biological.[5] (Not coincidentally, a new term of opprobrium, *pseudoscience,* entered the English vocabulary in the 1830s.) An Irishman writing in the 1860s noted that he would, "for convenience's sake, use the word 'science' in the sense which Englishmen so commonly give to it; as expressing physical and experimental science, to the exclusion of theological and metaphysical."[6] The new terminology signified a sharp break with the often explicit religious goals of natural philosophy. For Christians and non-Christians alike, a scientific explanation came to mean a naturalistic explanation.[7]

This change in nomenclature understandably alarmed and irritated groups such as theologians and linguists, who now found themselves outside the new boundaries of science—just at the time science was acquiring unprecedented cultural authority. Few persons felt this shift in status more acutely than Charles Hodge, the leading Protestant theologian in mid-nineteenth-century America. Hodge was no stranger to science. After finishing seminary, he had spent a winter in Philadelphia improving himself and attending lectures on anatomy and physiology at the medical school of the University of Pennsylvania, and he maintained a lifelong interest in scientific matters. He venerated "men of science who, while remaining faithful to their higher nature, have enlarged our knowledge of the wonderful works of God." However by the end of his career he was bitterly complaining that the very word *science* had become "more and more restricted to . . . the facts of nature or of the external world," that theology was losing its claim to scientific status, and that its practitioners were finding themselves increasingly regarded as objects of suspicion rather than beacons of light.[8]

The brilliant European-trained linguist William Dwight Whitney, brother of the geologist Josiah and master of many arcane languages, strongly resented the attempts of natural scientists to monopolize the term *science* for their own work, an exclusionary move that seemed by midcentury to be winning acceptance among "the community of scholars and cultivated men." In 1865 he went public with his complaint. "There is a growing disposition on the part of the devotees of physical studies—a class greatly and rapidly increasing in numbers and influence—to restrict the honorable title of science to those departments of knowledge which are founded on the immutable laws of material nature, and to deny the possibility of scientific method and scientific results where the main element of action is the varying and capricious will of man," he noted disapprovingly in the *North American Review.* "The name *science* admits no such restriction." The haughty astronomer Benjamin Gould, though personally friendly, especially irritated him. As Whitney complained to his brother, Gould "did not think to apologize for monopolizing the name 'science' to the *materialische* branches of knowledge."[9]

Although natural philosophers and men of science had long discussed the best means of teasing information out of nature, it was not until the latter part of the nineteenth century that they began claiming to possess a distinctive methodology, called "the scientific method." Scientists disagreed greatly about the actual content of this method, but by the early twentieth century it had become a staple of scientific apologetics, appearing in textbooks and the popular press alike. Rightly or wrongly, it seemed to give those of science a way of accessing knowledge that those of the cloth lacked.[10]

In his classic study *The Origins of Modern Science, 1300 to 1800* (1949), Herbert Butterfield famously claimed that "since the rise of Christianity . . . no landmark in history" has rivaled the Scientific Revolution in importance.[11] But for whom was the alleged revolution so important? Most Europeans could not read, and of those who could, only the most learned were fluent in Latin, the language of choice for natural philosophy. Even readers fluent in Latin could not always follow the reasoning of some of the leading philosophers of nature. When Francis Bacon sent King James I of England a copy of his *Novum organum* (1620), now regarded as one of the founding documents of modern science, the uncomprehending king likened it to "the peace of God, that passeth all understanding." Another acquaintance of Bacon's caustically noted "that a fool could not have written such a work, and a wise man would not."[12]

The founders of modern science often treated the common people with contempt. The great German astronomer Johannes Kepler, in dedicating his *Mysterium cosmographicum* (1596) to his noble patron, insisted that he wrote "for philosophers, not for pettifoggers, for kings, not shepherds." He dismissed "the majority of men" as too stupid and ignorant to appreciate his work. The English natural philosopher Isaac Newton reportedly told an acquaintance that "he designedly made his *Principia* abstruse" in order to "avoid being baited by smatterers in Mathematicks." A spokesman for the newly founded Royal Society in England celebrated the fellows' lack of "ambition to be cry'd up by the common Herd."[13] Given the lack of interest that most people showed in the mystifying worlds of natural philosophy and natural history, it may have been prudent of Kepler, Newton, and the founders of the Royal Society not to seek public approval. However there is no excuse for historians of science and religion to adopt a similarly condescending attitude toward the common people.[14]

Thanks to recent research on the history of popular science and popular Christianity, we are now in a good position to sketch out a new, populist narrative, indicating what the "vulgar"—the farmers and merchants, the homemakers and artisans—thought about the revolutionary scientific changes taking place around them.[15] The information at hand suggests that the new astronomy, the traditional centerpiece of the Scientific Revolution, attracted little notice and even less assent. Despite all of the subsequent attention heaped on Nicolaus Copernicus for dislodging the Earth from the center of the solar system and setting it in motion around the Sun, he won few converts in the period between the publication of his *De revolutionibus* (1543) and the end of the sixteenth century. A leading Copernican scholar has found only *ten,* though he may have missed one or two.[16] Even the intellectual elite of Europe virtually ignored the debate between geocentrists and heliocentrists before about 1615, the date Galileo Galilei used to mark "the beginning of the uproar against Copernicus." Most astronomers and theologians seemed to have reached agreement by the beginning of the eighteenth century, but large numbers of laypersons remained unpersuaded that they were whipping around the Sun at a ridiculously high speed.[17]

In focusing on Copernicus and other great men of science, we blind ourselves to the issues troubling the greatest number of believers, most of whom remained oblivious to the alleged theological implications of elite science. In many instances the public reacted to popularized versions of science and theology that trickled down from professional circles.

In other cases—Christian Science, creation science, and Native American science come readily to mind—various publics constructed their own alternative "science." While intellectuals wrestled with the theological ramifications of heliocentrism and the mechanical philosophy, the nature of force and matter, the manifestations of vitalism, the meaning of thermodynamics, relativity theory, quantum physics, and the implications of both positivism and scientific naturalism, the common people, to the extent that they paid any attention to science at all, concerned themselves largely with developments that impinged on their daily lives and self-understanding: monstrous births and meteorological abnormalities, phrenology and plural worlds, diseases, disasters, and descent from apes. As historians, we should go and do likewise.

A final observation: In writing about "science and religion," we have all too often focused our attention almost exclusively on Christianity, with perhaps an occasional glance at Jews and Muslims. At a time when religious ignorance and misunderstanding have lethal consequences, we need more than ever to avoid such provincialism. We must move the historical study of science and religion beyond its largely Christian base to include not only the other Abrahamic faiths but the religious traditions of Africa, Asia, and the Americas as well.

Notes

1. William R. Shea and Mariano Artigas, *Galileo in Rome: The Rise and Fall of a Troublesome Genius* (New York: Oxford University Press, 2004). See also David C. Lindberg, "Galileo, the Church, and Cosmos," in *When Science and Christianity Meet,* ed. Lindberg and Ronald L. Numbers (Chicago: University of Chicago Press, 2003), 33–60.

2. John L. Heilbron, *The Sun in the Church: Cathedrals as Solar Observatories* (Cambridge, Mass.: Harvard University Press, 1999), 3.

3. On the "complexity thesis," see part 2 of the current volume, as well as John Hedley Brooke, *Science and Religion: Some Historical Perspectives* (Cambridge: Cambridge University Press, 1991), and David C. Lindberg and Ronald L. Numbers, eds., *God and Nature: Historical Essays on the Encounter between Christianity and Science* (Berkeley: University of California Press, 1986).

4. John Winthrop, quoted in Charles Clark, "Science, Reason, and an Angry God: The Literature of an Earthquake," *New England Quarterly* 38 (1965): 353. On the history of "science," see especially Andrew Cunningham, "Getting the Game Right: Some Plain Words on the Identity and Invention of Science," *Studies in History and Philosophy of Science* 19 (1988): 365–89.

5. Ronald L. Numbers and Daniel P. Thurs, "The Scientific Idea," in *Encyclopedia of American Cultural and Intellectual History,* ed. Mary Kupiec Cayton and Peter W. Williams, 3 vols. (New York: Scribner's, 2001), 3:141–49.

6. [W. G. Ward], "Science, Prayer, Free Will, and Miracles," *Dublin Review* 8 (1867): 255.

7. Ronald L. Numbers, "Science without God: Natural Laws and Christian Beliefs," in Lindberg and Numbers, *When Science and Christianity Meet,* 265–85.

8. Ronald L. Numbers, "Charles Hodge and the Beauties and Deformities of Science," in John W. Stewart and James H. Moorhead, eds., *Charles Hodge Revisited: A Critical Appraisal of His Life and Work* (Grand Rapids, Mich.: Eerdmans, 2002), 77–102.

9. Stephen G. Alter, *William Dwight Whitney and the Science of Language* (Baltimore: Johns Hopkins University Press, 2005), 98–100, 139; William D. Whitney, "Is the Study of Language a Physical Science?" *North American Review* 101 (1865): 434–74.

10. Daniel Patrick Thurs, "Science in Popular Culture: Contested Meanings and Cultural Authority in America, 1832–1994" (Ph.D. dissertation, University of Wisconsin–Madison, 2004).

11. Herbert Butterfield, *The Origins of Modern Science, 1300–1800,* rev. ed. (1949; repr., London: Bell and Sons, 1957). See part 3 of the current volume for a discussion of Butterfield's thesis.

12. Lisa Jardine and Alan Stewart, *Hostage to Fortune: The Troubled Life of Francis Bacon* (New York: Hill and Wang, 1999), 439.

13. Johannes Kepler, *Mysterium Cosmographicum: The Secret of the Universe,* trans. A. M. Duncan (1596; New York: Abaris, 1981), 55–57; Frank E. Manuel, *Isaac Newton, Historian* (Cambridge, Mass.: Harvard University Press, 1963), 260; and Joseph Glanville, quoted in Larry Stewart, *The Rise of Public Science: Rhetoric, Technology, and Natural Philosophy in Newtonian Britain, 1660–1750* (Cambridge: Cambridge University Press, 1992), xxviii. I am grateful to Robert S. Westman for bringing the Kepler quotation to my attention.

14. One discovers little about popular views in Brooke's influential *Science and Religion* or—I regret to say—in the two collections I have coedited with Lindberg, *God and Nature* and *When Science and Christianity Meet.* Slightly more inclusive is *The History of Science and Religion in the Western Tradition: An Encyclopedia,* ed. Gary B. Ferngren, with the assistance of Edward J. Larson, Darrel W. Amundsen, and Anne-Marie E. Nakhla (New York: Garland, 2000).

15. This account of popular science and religion is based on Ronald L. Numbers, "Science and Christianity among the People: A Vulgar History," in *Modern Christianity to 1900,* ed. Amanda Porterfield, vol. 6 of *A People's History of Christianity,* ed. Denis R. Janz, 7 vols. (Minneapolis: Augsburg Fortress, 2007).

16. Robert S. Westman, "The Astronomer's Role in the Sixteenth Century: A Preliminary Study," *History of Science* 18 (1980): 105–47.

17. Maurice A. Finocchiaro, *Retrying Galileo, 1633–1992* (Berkeley: University of California Press, 2005), 72. For an excellent account of popular Copernicanism, see Rienk Vermij, *The Calvinist Copernicans: The Reception of the New Astronomy in the Dutch Republic, 1575–1750* (Amsterdam: Koninklijke Nederlandse Akademie van Wetenschappen, 2002).

Rejoinder to Numbers and Larson

William R. Shea

Ronald Numbers and Edward Larson have been both perceptive and kind. I am grateful for their responses, each of which points out some ambiguities in my essay and offers the opportunity to sort out and clarify the issues. I shall begin with Numbers's suggestions and then move on to Larson's comments.

I am pleased that Numbers should stress, in a personal and therefore more interesting way, the fact that it is generally impossible to know the religious orientation of recent historians writing about the relationships (I accept Numbers's plural) between science and religion. I see this as a clean bill of health for the profession, for if there is anything that historians want to avoid, it is the projection of their own concerns onto the past, thereby unwittingly collapsing the distance between past and present. To recuperate a lost world of thought, we need a blend of faith and agnosticism. We have to believe that we can enter that world, and we have to be skeptical lest we claim to have found there what we ourselves have brought along.

The categories we use to explore the past have a way of shifting out of focus when we move backwards in time. As Numbers intimates, *science* and *religion* do not inhabit a serene noetic heaven but live in the hurly-burly here below. Even if we could agree on suitably general definitions, say, "belief in a personal God or gods entitled to obedience and worship" for religion and "knowledge of the natural world, based on controlled experiment, and arrived at by the use of reason, not revelation" for science, we would still want to know by whom these views were stated and in whose minds they clashed, were harmonized, or just piled up side by side. For the historian (I dare not speak for the philosopher), science and religion as such cannot interact. They enter the real world only insofar as they are defended by humans and, as David C. Lindberg nicely puts it, "When flesh and blood make an appearance,

From *Historically Speaking* 7 (November/December 2005)

we are apt to find that personal interest and political ambition are as important as ideological stance."[1]

Unless we subscribe to the dogma of historical inevitability (fatalism in high gear), we cannot but be impressed by the way the outcome of Galileo's trial was influenced by local circumstances, such as the fact that everyone involved, without exception, acknowledged the authority of the Bible and that no one, least of all Galileo, defended the notion that a stable society could be built on the democratic principles that we take for granted in the twenty-first century. Were a historically sensitive movie director to choose to make a film about Galileo there would be no clearly recognizable "good guys" and "bad guys." No firm would underwrite the cost of production. It is well known that commercial films sacrifice historical authenticity for broad audience appeal, simplifying the complex patterns of the past and telling the public what it wants to hear.

Which brings me to Numbers's urgent request that we cast our nets wider and include in our research not only the great and the little men of science but the common people. Today they get their ideas about science by watching TV, surfing the Web, or reading popular magazines. I would love to know how they went about it when the average person could neither read nor write, as in Galileo's day. I find Numbers's proposal fascinating, and I look forward to the kind of fish that the new nets will haul in, provided the meshes are smaller and tighter. In any case I know of no one better prepared than Numbers to sail on the high seas of popular culture. I cannot resist a *caveat,* however. When he writes that "the founders of modern science often treated the common people with contempt," I am compelled to tear a leaf out of his own book and recall that we generalize at our peril. I shall not attempt to contextualize Kepler's remark, but I would like to quote the very first lines of the *Two New Sciences* that Galileo published in 1638 where his spokesman, Salviati, declares, "The constant activity which you Venetians display in your famous arsenal suggests to the studious mind a large field for investigation, especially that part of the work which involves mechanics; for in this department all types of instruments and machines are constantly being constructed by many artisans, among whom there are some who, partly by inherited experience and partly by their own observations, have become highly expert and whose reasoning is of the finest."[2] The common artisan could hardly have wished for more.

Historians occasionally take a sidelong glance at sociology and anthropology. Numbers is not only willing to look closely at their fields; he is actually packing his bags for a long voyage into realms where ordinary mortals hold deep, if sometimes only partly articulated, beliefs but have not come to terms with the new science. I hope to learn much from his trip, and I am equally interested in the second one that he plans to take beyond the shores of the three great monotheistic religions. I wish I could accompany him in the unexplored regions where the traditional religions of Africa, Asia, and the Americas met or, perhaps, just glided past modern science. If I can't go, he will find me cheering on the shore as he and his fleet cast their moorings.

I now turn, my two feet on the solid, albeit uneven, ground of seventeenth-century science (the only one where I can stand up comfortably) to face Edward Larson's "pacifist" defense of *warfare* as a suitable metaphor to describe some incidents that he has studied and that are a long way from those that involved Galileo and Kepler. But let me first make a small inroad into semantics. Larson speaks of "March Madness" as a suggestive metaphor for what happens on college campuses during the annual basketball tournament. The incessant dribbling "does not actually drive basketball fans *mad* in the clinical sense of that term." I agree entirely, but may I suggest that *madness* here does not have the sense of "having a disordered mind" but rather of "being wildly excited." I know colleagues, even historians, who avow that they are *mad* about football or hockey. When I hear this, I rejoice that they are still capable of enthusiasm. Their declaration reinforces my conviction that they have not gone to seed or become boring bookworms. To be *mad,* in this sense, is a good thing. It would be much more difficult to give a positive meaning to *warfare.* Since the nineteenth century it has reinforced the sense of discord between religion and science and thrown historians into adverse camps. I feel strongly, in the light of what knowledge I have been able to acquire about the Galileo affair, that labels such as *warfare* and *conflict* hinder us in our task of uncovering the otherness of the past and translating it into a language accessible to our contemporaries. The military and belligerent connotations that travel in the company of *warfare* crowd out other linguistic passengers who carry finer and more varied luggage.

I do not wish to downplay the seriousness of freedom of research but to plead for the recognition that phrases such as *battle line* (a close friend of *warfare*) are misleading in the case of Galileo. There was no army of scientists, determined to overlook the Bible or interpret it allegorically, facing a cohort of geocentric theologians, prepared to save biblical inerrancy at all cost. What we find is a more complex situation: Many scientists were respectful of the caution urged by Bellarmine, and a significant number of clergymen accepted Galileo's exegetical principles. The "struggle" (I happily go this far) was located *within* the Church as much as *within* science. This is why I feel uneasy with the compliment that Larson apparently pays me when he writes:

"I readily accept the historical significance that he ascribes to the Roman Catholic Church's *persecution* of the Italian scientist." But my point is that *persecution* is a blinker word that we must remove from our glossary if we are to see what really happened. I must have expressed myself very badly if I failed to convince Larson, one of the finest historians I know. The problem may also lie in the fact that Larson deals with much more recent events than I do and that he stresses the bond between past and present whereas I live in dread of anachronism. I find Larson's most recent book, *Evolution: The Remarkable History of a Scientific Theory,* fascinating, and I warmly recommend it to anyone who is interested in knowing how we got from Darwin to the creationists.[3] There is much mudslinging and little understanding between evolutionists and those who believe that *Homo sapiens* was created a few thousand years ago. Why this should be the case is obvious to people living in America but puzzling elsewhere.

If I understand Larson correctly, the issue is less a matter of sound science challenging genuine Christianity than a question of pseudoscience taunting people passionately attached to traditional values. In some cases, as in the nineteenth century when run-of the-mill believers were under attack by Francis Galton, it may have been a question of survival. Galton, who felt himself English to the core, proclaimed that it was "scientifically" and, hence, "morally" necessary to improve the race by siring good offspring (he managed fourteen in and out of wedlock) and curtailing the fertility of inferior races, including blacks and lesser white breeds such as the Irish, the Scots, and the Welsh.[4] Galton did some useful research, and Larson bends over to be fair to him, but it would be more than lenient to call him *mad* in the sportive way the word is applied to basketball fans.[5]

If Galton's eugenics was a caricature of science, a reader who is not acquainted with the squabble (I almost wrote the *front*) might consider that the twentieth-century evangelist Billy Sunday, who "jumped, kicked, and slid across the stage" in a perfectly rehearsed and choreographic way, was putting on a vaudeville show rather than delivering the Christian message.[6] But we are now back where we began. What is *science,* and what is *religion?* Galton's ideas were taken seriously by his cousin Charles Darwin, and Billy Sunday's theatrics won him invitations to preach from a broad spectrum of Protestant churches.

Notes

1. David C. Lindberg, "Galileo, The Church, and the Cosmos" in *When Science and Christianity Meet,* ed. Lindberg and Ronald L. Numbers (Chicago: University of Chicago Press, 2003), 57.

2. Salviati, quoted in Galileo Galilei, *Two New Sciences,* trans. Henry Crew and Alfonso De Salvio (New York: Macmillan, 1914), 1. Slightly revised translation.

3. Another excellent account is Karl W. Giberson and Donald A. Yerxa, *Species of Origins: America's Search for a Creation Story* (Lanham, Md.: Rowman & Littlefield, 2002).

4. It is a sobering thought that I would not be writing these lines had Galton had his way.

5. On eugenics, see the chilling accounts of Daniel J. Kevles, *In the Name of Eugenics* (New York: Knopf, 1985) and André Pichot, *La société pure de Darwin à Hitler* (Paris: Flammarion, 2000).

6. Edward Larson, *Evolution: The Remarkable History of a Scientific Theory* (New York: Modern Library, 2004), 201.

PART 2

Complexity and the History of Science and Religion

Science, Religion, and Historical Complexity

John Hedley Brooke

Few discourses have been as riven with prejudice and polemical intentions as those concerning the mutual bearings of science and religion. The spectacular example of Richard Dawkins's antireligious mission in *The God Delusion* and the scathing reviews it has provoked in both the *London* and *New York Review of Books* testify to ongoing battles and high public interest.[1] In the popular mind, science and religion are still engaged in a centuries-long war. Yet historians, drawing on longer perspectives and a rich diversity of interpretation, have contested the popular claim that science and religion are—and always have been—inevitably in conflict. Some of these historians, motivated by religious sympathies to "set the record straight," have perhaps gone too far in the other direction. In contrast to the conflict thesis, a meta-narrative of peace (or at least the potential for peace) has repeatedly found expression, as in a recent essay of David C. Lindberg, one of our finest historians of science: "In those not infrequent cases where Christianity and science have attempted to occupy the same intellectual ground, the historical actors have generally preferred peace to warfare, compromise to confrontation, and have found means—through compromise, accommodation, clarification, re-interpretation, revision and the identification of outright error—of negotiating a state of peaceful coexistence."[2]

Through the exposure of historical myths, inscribed for example in popular accounts of the Galileo affair or of the ignominious defeat of Bishop Samuel Wilberforce at the hands of Darwin's disciple Thomas H. Huxley, more balanced views have become possible.[3] Particularly for early modern Europe, when the concerns of natural philosophy and theology were sometimes fused together, a fascinating picture emerges of scientific activity grounded in, and justified by, theological considerations.[4] In many strands of

From *Historically Speaking* 8 (May/June 2007)

natural theology, the scientific disclosure of beauty, harmony, and *apparent* design mediated between scientific and religious interests, as when Isaac Newton insisted that the beauty of the solar system could *only* have originated in the "counsel" of a divine being.[5]

One consequence of serious historical research is a growing suspicion of metanarratives and an insistence on historical complexity. As one who must plead guilty to fueling such suspicion, I have myself been identified as a purveyor of what has come to be known as the "complexity thesis." I shall raise the question later whether this is a satisfactory appellation.

But it may be helpful first to indicate some of the reasons why one cannot suppress reference to complexity. The meaning of the word *science* has changed over time. The word *scientist* was not coined until the 1830s. And the word *religion* draws its modern connotations from Enlightenment ambitions to impose comparative structures on the study of different societies and their rituals.[6] Such facts might not be sufficient to torpedo a philosophical quest for normativity based on redefinition, but other considerations may do so. The views of seminal thinkers may themselves change with time and circumstance. Even a simple diachronic model, such as Charles Darwin's shift from a Christian to a deistic to an agnostic phase, meets the complication (in his own words) that "my judgment often fluctuates."[7] Even an apparently unproblematic prescription, such as a characteristically "inductive" methodology for the sciences, turns out to conceal deep rifts of meaning, as Laura Snyder has shown in her brilliant analysis of the dispute between William Whewell and John Stuart Mill.[8] Again, those who like to speak of the relations between science and religion need to be reminded how easily, but misleadingly, these become singularized, hypostatized terms, concealing a diversity of social practices among the sciences (plural) and among the major faith traditions (plural). Moreover, controversies supposedly between science and religion often turn out to be between competing scientific theories, in which religious interests may be at stake, or between rival religious groups seeking to appropriate the authority of science for their particular cause. Very different sets of implications for religious belief may be claimed for one and the same scientific theory, as is abundantly clear from the many affirmative as well as dismissive responses to Darwin's theory of evolution.[9]

Similarly, one and the same religious doctrine has, in different contexts, been both conducive and stultifying to scientific initiative. Seventeenth-century Puritan ministers, believing as they did in the doctrine of the "fall," felt that the very notion of rational, scientific inquiry was presumptuous.[10] But, as Peter Harrison has recently emphasized, a desire to recover at least some of the pristine knowledge of Adam, lost at the fall, framed many discussions of scientific methodology in that same period—Francis Bacon affording a prime example.[11] Whereas we automatically think of biblical literalism as obstructive to scientific freedom, Harrison has shown how, in earlier periods, it could assist the demystification of the book of nature.[12] A doctrine of Creation could underscore belief in the order and intelligibility of nature, as it clearly did for Newton; but in later periods the doctrine was easily vulgarized to prohibit biological evolution. How such connections are made must also depend on the epistemological status of scientific theories and the ontological status of the theoretical entities they posit. Whereas popular texts in the philosophy of science often suggest that a choice has to be made between realist and instrumentalist accounts, the historian is apt to see both philosophies at work according to time and place. This is not to diminish the importance of the distinction, which was fundamental in early debates over the Copernican system. If the heliocentric theory were merely a mathematical model, the ecclesiastical authorities had little to fear; if, however, it purported to represent the physical structure of the cosmos, as Galileo preferred, then there were pressing issues of biblical reinterpretation. The complexity runs deeper still because the historian sees that some sciences have been destined to be more instrumentalist than others, especially when direct verification, as in many cosmological models, has been elusive.

Amid such complexity, efforts have been made to clear the air, as when the late Stephen Jay Gould suggested that the *magisteria* of science and religion should not be permitted to overlap, the former having jurisdiction over the facts of nature, the latter over moral values.[13] This partitioning has much to commend it as a default position, but it fails for several reasons. It is reductive both of the explanatory ambitions of the sciences and of the truth-seeking quest of most religions. Second, it fails to do justice to the fact that propositions derived from religious belief may indeed have nothing to do with scientific propositions at one level but be profoundly affecting or affected at another. Francis Bacon has sometimes been hailed for his secularity because he cautioned against mixing theistic reference with efficient causes when explaining specific natural phenomena. But, on other levels, Bacon saw scientific inquiry as a religious duty, conducive to the virtue of humility and, as mentioned above, restorative of a lost knowledge and dominion originally granted to Adam. When questions of patronage rather than epistemology are introduced, there can be other twists, as when John Heilbron showed just how much physics was done *inside* churches. In a passage that has been cited in the previous forum in this volume, Heilbron wrote that "the Roman

Catholic church gave more financial and social support to the study of astronomy for over six centuries, from the recovery of ancient learning during the late Middle Ages into the Enlightenment, than any other, and, probably, all other, institutions."[14]

Yet seemingly plausible attempts to correlate distinctive religious traditions with receptivity, or the lack of it, toward specific scientific theories are also liable to come unstuck—for the reasons that David N. Livingstone has so persuasively given when contrasting different sites of knowledge production, resistance, and assimilation.[15] As he has demonstrated, even within the same theological tradition—Presbyterianism—the manner in which Darwinian ideas were evaluated and the timing and trajectory of their assimilation differed markedly in Princeton compared with Belfast and were different again in Edinburgh. In New Zealand Darwinism was a welcome resource for justifying colonial extermination of the Maori, while in the southern states of America, Darwin's insistence (and that of Alfred Russell Wallace also) on the unity of the human race was not always welcome in a culture of racial hierarchy. Livingstone has placed special emphasis on local events in shaping predispositions, such as John Tyndall's notorious presidential address given in Belfast in 1874 before the British Association for the Advancement of Science. Tyndall's streamlining of a triumphalist history, in which Darwinism was part and parcel of a bid to wrench cosmology from the theologians, so alienated sensitive members of his audience that in Northern Ireland Darwin's theory was especially liable to be associated with materialism and atheism.[16]

As historians map their local and manifold contingencies, so the topographies of complexity increasingly interpose. Consequently, I stand by my own formula: There is no such thing as *the* relationship between science and religion. When I wrote in 1991 that the lesson of history is its complexity, I was targeting the facile use of historical myths and anecdotes both by religious apologists and their detractors to defend their supposedly normative but often partisan, simplistic models.[17] It had not occurred to me that my insistence on historical richness, complexity, and the need for multiple stories would be elevated by some commentators into a "complexity *thesis,*" to be placed, presumably, alongside the "conflict thesis," the "Merton thesis," the "Foster thesis," the "Lynn White thesis," and other well-known correlations between religious values and their expression in the sciences.[18] I would prefer to say (unsurprisingly!) that complexity is a reality, not a thesis, and that, instead of being placed alongside other theses, it should be permitted, at least in the first instance, to function as critique.

Two main criticisms have been leveled against what some might see as a reveling in complexity for its own sake. Judged from within a constituency where the status of rival scientific and religious beliefs is a deadly serious matter—as among young-Earth creationists eager to demonize Darwin—any historically grounded attempt to show the wisdom of other options is liable to miss the stark, present reality of a conflict having many social and political dimensions. This, it seems to me, would be a valid objection if the appeal to complexity were meant to eliminate conflict by a sleight of hand. But there is nothing in such an appeal that requires us to expunge conflict from our gaze when its reality stares us in the face. If there is the prospect of edification from an appeal to complexity, it is because there is usually more on offer from within specific religious traditions than their most extreme modern representatives either know or wish to know.

A second objection, put baldly, is that complexity is not enough. One's students or one's readers need something less protean. The local must surely be balanced by the global. Otherwise there will be a tendency to throw up one's hands in despair at the multiplicity and the fragmentation. Does the historian not have a duty to detect the patterns, however difficult their discernment might be? I think this would be a cogent objection if advocates of complexity had reneged on their obligation. It is, however, perfectly possible to look for patterns behind the complexity. And if they can be perceived and articulated only at a metalevel, they may still sustain generalities of a kind. Perhaps two examples from recent scholarship on Reformation religion and the sciences will help. One is concerned with Lutheran, the other with Calvinist attitudes.

Strong claims have been made for Lutheran advocacy of Copernican astronomy, even to the point of suggesting that one might pinpoint a specifically Lutheran natural philosophy.[19] In their respective contributions to a European Science Foundation project on religious values and the rise of science, Charlotte Methuen and Anne-Charlott Trepp both feel obliged to deconstruct this supposedly characteristic Lutheran science. Methuen wishes to show the diversity of Lutheran sensibilities during the period of confessional formation, Trepp that the study of nature was not a derivative of a particular religious creed but, as in the case of her exemplar Johann Rist, might supply a kind of reassurance on matters of salvation that the doctrine of justification by faith had failed to provide.[20] The sophistication of their respective inquiries does not, however, preclude a metalevel patterning. Methuen suggests that "those taking new interest in observing the natural world might form a subset of several of the increasingly disparate theological camps of the sixteenth century which transcend confessional boundaries." Trepp makes a similar move, insisting that seventeenth-century Lutheranism was characterized less by rigid orthodoxy than by an "internal process of differentiation and pluralization that also ultimately included orthodoxy." Hence her metalevel pattern: It was precisely these pluralizing forces that were decisive for the interpretation and study of nature as a form of religious practice in the seventeenth century, and it was "precisely the theologians who professed the 'true Christianity' and distanced themselves from the disputational theology of the universities . . . who displayed a particular interest in nature." Asking whether there was a distinctively Calvinist interest in nature in the Netherlands, Kenneth Howell finds it necessary to undertake a comparable deconstruction. Including in his account the contrasting natural philosophies of two Dutch microscopists, Antoni Leeuwenhoek and Jan Swammerdam, both of whom adhered to a reformed faith, Howell has to conclude that "the debates among Dutch Calvinists from the 1650s onward show quite clearly the impossibility of linking too closely the acceptance of Cartesianism and/or Copernicanism with some propensities resident in Calvinist theology." Indeed "the followers of Calvin stood on both sides of the philosophical divide." But not all patterning is thereby erased. Howell can still say that "almost all Calvinists rejected scholastic philosophy because of its close ties with Romanism."[21]

None of these recent studies rules out other metalevel possibilities, such as the shaping of an individual's science by religious preconceptions or the challenge to sacred texts from scientific innovation. In a collection of case studies devised to explore the ways in which religious (and antireligious) beliefs might have informed the content of scientific theories, one contributor, Bernard Lightman, nicely captures the metalevel possibilities when he argues that the well-known Victorian conflict was not between science and religion but was, more subtly, the result of dissonance among the harmonizers. In part, the quarrel was about how best to preserve the *consonance* of scientific with religious thinking.[22] Of these more subtle patterns there are doubtless many. Is it not wiser, for example, to think of the challenging scientific innovations as divisive within religious communities rather than as undifferentiated cognitive threats? In his recent study of *Quakers, Jews, and Science,* Geoffrey Cantor has observed how much more historical work needs to be done in exploring the *intrareligious* dialogues that have been precipitated by innovative, intrusive science.[23]

On the basis of historical examples, it would be difficult to deny that one and the same scientific theory can usually be exploited for both theistic and atheistic purposes. Instead of capitulating to Dawkins's insinuation that one cannot be a real scientist without being an atheist, should we not also acknowledge a more delicate pattern? That is, when scientists have been reverent, as many have, they will often be found among the more liberal or idiosyncratic exponents of their faith tradition, not least because the skeptical approach characteristic of their scientific outlook can be a solvent for dogmatic orthodoxies.[24] This pattern is not infallible, of course, and there has been scope for other correlations, such as the seemingly greater appeal of ultraconservative religious positions to physical and engineering scientists than to those immersed in the life sciences, arising perhaps from a "unique solution" mentality more characteristic of the former.

The awe and wonder sometimes expressed by scientists as they marvel at the intricacies and splendor of nature make it difficult to propose a simple correlation between science and secularity. There is, however, the more subtle pattern identified by Peter Burke, who once astutely observed that scientists have been destructive of the sacred "in spite of themselves."[25] No account of the receptivity of religious groups toward scientific movements can work unless it recognizes a recurring gulf that so often separates popular reaction and sophisticated academic judgment. Here is the Unitarian minister Francis Ellingwood Abbot writing to Charles Darwin on August 20, 1871: "If I rightly understand your great theory of the origin of species, it contains nothing *inconsistent* with the most deep and tender religious feeling. It certainly conflicts with the popular notion of God, but it seems to me to harmonize thoroughly with the enlightened ideas concerning him held by all highly cultured minds of today . . . and for one I feel that you have done a vast service to true religion by your labors."[26] Darwin himself was characteristically reticent when confronted with such opinion; but that gulf between lay and learned has to feature in any big picture. Galileo deemed the mobility of the Earth to be a proposition far beyond the comprehension of the common people. And in our own day, surveys of the religious beliefs of American scientists suggest a higher degree of skepticism or atheism among those described by Edward J. Larson as occupying the "higher echelons of the profession."[27]

It is not that the recognition of complexity permits no patterns. It is, rather, that the patterns themselves are more intricate. How often have the Christian churches run into trouble because their representatives have embraced uncritically a voguish piece of science, only to be left stranded when it proves defective? How often have religious apologists run into trouble (scientists, too, for that matter) by overburdening the sciences with religious freight?[28] How often has an apologetic appeal to a deity to close what turn out to be only transient gaps in scientific understanding eventuated in embarrassing mistakes? How often, finally, have the contenders in debates over theism and naturalism failed to recognize that their (and their opponents') contentions may be embedded in worldviews whose presuppositions and ramifications may not be immediately visible?[29] It is surprising, for example, how many scientific missionaries for atheism have imagined that all they need to do to vindicate their position is to show that a naturalistic explanation is *possible* for a phenomenon that has been imbued with religious significance. There is blindness here to the fact that a classical theistic worldview can accommodate the integrity of naturalistic explanation without sacrificing the complementarity of a primary causality, a primary dependence on the sustaining role of a transcendent power. The postulate of dependency may have no bearing on the practice and content of the sciences, but that has not itself been sufficient to prevent its affirmation by those for whom it has religious meaning.

Notes

1. Richard Dawkins, *The God Delusion* (New York: Bantam, 2006); Terry Eagleton, "Lunging, Flailing, Mispunching," *London Review of Books,* October 19, 2006; and H. Allen Orr, "A Mission to Convert," *New York Review of Books,* January 11, 2007.

2. David C. Lindberg, "Of War and Peace," *Science & Theology News* (March 2006): 33.

3. For correctives on the Galileo affair, see William Shea and Mariano Artigas, *Galileo Observed: Science and the Politics of Belief* (Sagamore Beach, Mass.: Science History, 2006) and Ernan McMullin, ed., *The Church and Galileo* (South Bend, Ind.: University of Notre Dame Press, 2005). Revisionist scholarship on the Wilberforce-Huxley debate can be approached through Frank James, "An 'Open Clash between Science and the Church'?: Wilberforce, Huxley and Hooker on Darwin at the British Association, Oxford, 1860," in *Science and Beliefs: From Natural Philosophy to Natural Science, 1700–1900,* ed. David Knight and Matthew Eddy (Aldershot, Hampshire, U.K.: Ashgate, 2005), 171–93, and John Brooke, "The Wilberforce-Huxley Debate: Why Did It Happen?" *Science and Christian Belief* 13 (2001): 127–41.

4. Amos Funkenstein, *Theology and the Scientific Imagination from the Middle Ages to the Seventeenth Century* (Princeton, N.J.: Princeton University Press, 1986).

5. For an introduction to the links between natural theology and the natural sciences, see John Hedley Brooke and Geoffrey Cantor, *Reconstructing Nature: The Engagement of Science and Religion* (Edinburgh: Clark, 1998), chaps. 5–7.

6. Peter Harrison, *"Religion" and the Religions in the English Enlightenment* (Cambridge: Cambridge University Press, 1990); Geoffrey Cantor and Chris Kenny, "Barbour's Four-Fold Way: Problems with His Taxonomy of Science-Religion Relationships," *Zygon* 36 (2001): 763–79.

7. Francis Darwin, ed., *The Life and Letters of Charles Darwin,* 3 vols. (London: Murray, 1887), 1:304.

8. Laura Snyder, *Reforming Philosophy: A Victorian Debate on Science and Society* (Chicago: University of Chicago Press, 2006).

9. James Moore, *The Post Darwinian Controversies* (Cambridge: Cambridge University Press, 1979); Jon Roberts, *Darwinism and the Divine in America* (Madison: University of Wisconsin Press, 1988); David Livingstone, *Darwin's Forgotten Defenders* (Grand Rapids, Mich.: Eerdmans, 1987); and John Durant, ed., *Darwinism and Divinity* (Oxford: Blackwell, 1985).

10. John Morgan, "The Puritan Thesis Revisited," in *Evangelicals and Science in Historical Perspective,* ed. David N. Livingstone, D. G. Hart, and Mark Noll (New York: Oxford University Press, 1999), 43–74.

11. Peter Harrison, *The Fall of Man and the Foundations of Modern Science* (Cambridge: Cambridge University Press, 2007).

12. Peter Harrison, *The Bible, Protestantism, and the Rise of Natural Science* (Cambridge: Cambridge University Press, 1998).

13. Stephen Jay Gould, *Rocks of Ages: Science and Religion in the Fullness of Life* (New York: Ballantine, 1999).

14. John L. Heilbron, *The Sun in the Church: Cathedrals as Solar Observatories* (Cambridge, Mass.: Harvard University Press, 1999), 3; Ronald Numbers, "Comments on William Shea," *Historically Speaking* 7 (November/December 2005): 11.

15. David N. Livingstone, *Putting Science in Its Place: Geographies of Scientific Knowledge* (Chicago: University of Chicago Press, 2003).

16. David N. Livingstone, "Darwinism and Calvinism: The Belfast-Princeton Connection," *Isis* 83 (1992): 408–28, and "Science, Region, and Religion: The Reception of Darwinism in Princeton, Belfast, and Edinburgh," in *Disseminating Darwinism: The Role of Place, Race, Religion, and Gender,* ed. Ronald Numbers and John Stenhouse (Cambridge: Cambridge University Press, 1999), 7–38.

17. John Hedley Brooke, *Science and Religion: Some Historical Perspectives* (Cambridge: Cambridge University Press, 1991), 321.

18. David Wilson, "The Historiography of Science and Religion," in *The History of Science and Religion in the Western Tradition: An Encyclopedia,* ed. Gary Ferngren, with the assistance of Edward J. Larson, Darrel W. Amundsen, and Anne-Marie E. Nakhla (New York: Garland, 2000), 3–11, and Stephen Weldon, "The Social Construction of Science," in Ferngren, *History of Science and Religion,* 220–22.

19. For the special role of Lutherans in promoting Copernican astronomy, see Peter Barker, "The Lutheran Contribution to the Astronomical Revolution: Science and Religion in the Sixteenth Century," in *Religious Values and the Rise of Science in Europe,* ed. John Brooke and Ekmeleddin Ihsanoglu (Istanbul, Turkey: Research Centre for Islamic History, Art, and Culture, 2005), 31–62.

20. Charlotte Methuen, "On the Problem of Defining Lutheran Natural Philosophy," in Brooke and Ihsanoglu, *Religious Values,* 63–80; Anne-Charlott Trepp, "'Nature' as Religious Practice in Seventeenth-Century Germany," in Brooke and Ihsanoglu, *Religious Values,* 81–110.

21. Kenneth Howell, "Styles of Science, Calvinism and the Common Good in the Early Dutch Republic," in Brooke and Ihsanoglu, *Religious Values,* 111–30.

22. Bernard Lightman, "Victorian Sciences and Religions: Discordant Harmonies," *Osiris* 16 (2001): 343–66.

23. Geoffrey Cantor, *Quakers, Jews and Science* (Oxford: Oxford University Press, 2006), 356–57.

24. This particular pattern is indicated in Brooke, *Science and Religion,* 44–45.

25. Peter Burke, "Religion and Secularisation," in *The New Cambridge Modern History,* ed. Burke, vol. 13 (Cambridge: Cambridge University Press, 1979), 293–317, esp. 303.

26. I am indebted to Alison Pearn of the Cambridge Darwin Correspondence Project for this example.

27. Edward J. Larson, "Response to William Shea," this volume.

28. For complementary perspectives here, see John Dillenberger, *Protestant Thought and Natural Science* (New York: Collins, 1961), and Michael Buckley, *At the Origins of Modern Atheism* (New Haven, Conn.: Yale University Press, 1987).

29. For topical reflection on the distinction between Darwinian theory and Darwinism as a worldview, see Michael Ruse, *The Evolution-Creation Struggle* (Cambridge, Mass.: Harvard University Press, 2005).

Reply to Brooke

William R. Shea

Was science, which is based on reason and experiment, bound to clash with religion, which relies on authority and dogma? Why did the Church attempt to silence Galileo? And why did it try to shout Darwin down? These are some of the questions that John Brooke, the first incumbent of the Andreas Idreos Chair of Science and Religion at the University of Oxford, was asked not only by his students but by the large audiences that attended his lectures in various parts of the world. His answer was invariably, "let's look at the facts." And the facts have a way of being many sided.

Science and religion have found it difficult to reconcile their different viewpoints on two major issues. One is the structure of the world; the other the origin of man. The first is associated with the name of Galileo; the second with Darwin. The difference, and hence the complexity, is that in the seventeenth century the belief that there is a divine order in the universe was still secure in the minds of Galileo and his contemporaries. They might have been busy exploring the mechanical order, but they still took the old order of numinous truth for granted. Copernicus, who was a canon of the cathedral of his diocese in Poland, was convinced that the motion of the Earth was not incompatible with the Christian faith, and he dedicated his famous work *On the Revolutions of the Heavenly Spheres* to Pope Paul III. The German astronomer Johannes Kepler, who discovered the laws that describe how planets move around the Sun, wanted to become a pastor in the Lutheran Church, and he saw himself as an interpreter of God's works in nature. The man who thought of universal gravitation, Isaac Newton, spent more time commenting on scripture than he did working on problems of physics and mathematics. For these pioneers of the Scientific Revolution, the Bible was

From *Historically Speaking* 8 (May/June 2007)

not only divinely inspired but every statement had a spiritual meaning. Their problem was to find the real meaning, and they were convinced that the new science would assist them in that task. If scripture taught how to go to heaven, science taught how the heavens go. There was no need to deny the first lesson when the second was imparted.

The snag is that the scientific lesson was not always as convincing as it was exciting. Galileo was never able to offer a proof that the Earth moves around the Sun, although he was very clever at showing that the objections raised by his opponents begged the question. He had little patience with the qualms of the Catholics, who smarted under Protestant accusations that they neglected scripture, and reacted by adopting a fundamentalist stance. But if theologians went overboard in claiming revealed status for the traditional geocentric view, Galileo did not hesitate to ask for faith in his heliocentric hypothesis. Both sides were dogmatic and, unconsciously perhaps, more nervous than their hectoring would suggest. They shared, however, the idea that the Bible is God's Word in a sense that is no longer familiar to the twenty-first century.

When we hear the Bible described as inspired, we think of a book that contains, among other material, records of religious experiences that are capable to some degree of stimulating our own religious sensitivity. But the theologians that Galileo met in Italy maintained that any apparently factual statement in the Bible was to be regarded as such unless the contrary was proven. Since the Sun is said to rise and set in several passages of scripture, it seemed that the motion of the Sun was an astronomical fact. Until science could show that this is merely how things appear, there was no reason to take the risky step of declaring that the Bible had to be reinterpreted. Of course Copernicanism could be taught as a speculative theory and used to make computations. Should it one day be confirmed, then, and only then, would the Church have to recognize that it had unwarrantedly given a common-sense utterance a factual meaning that the scripture did not possess. But Galileo rushed where angels would have feared to tread, and the outcome was the clash that left him with the aura of a martyr and Pope Urban VIII with a posthumous black eye. The opposition between science and religion was not written in the stars, and if Galileo had been less arrogant and Urban VIII less proud, history might have followed a different course. But there was a deeper underlying issue. When Galileo insisted that God was the author of both nature and scripture, he implied that theologians would henceforth have to rely on scientists to interpret biblical statements about natural events. Many saw this as a blow to their authority, and their concern was heightened when a hundred years later German scholars began to query the notion that the

Bible was inspired from cover to cover. This time the doctrine of verbal inspiration became a stumbling block for Catholics and Protestants alike. The storm broke in England in the nineteenth century, and in this context Darwin's theory of evolution added more than mere disturbance.

Although some members of the clergy declared that evolution was not incompatible with the doctrine of creation, they could only do so by discarding the assumption that the Bible was a theological textbook composed and dictated *verbatim* by Almighty God. To come to terms with historical scholarship and natural science, they had to embrace the view that the invulnerability of scripture was an idol. In other words, they had to accept that they were indeed beholden to science. In this sense science won the day, but this did not lead, as many humanists fervently hoped, to the demise of religion. Bibliolatry came to be seen by believers themselves as a betrayal of Christianity, and a new generation approached the Bible not as a font of intellectual certainty but as a source of heart-awakening experiences about the human predicament.

If the story were to end here, all would be sweetness and light. But even when the problem of biblical interpretation was resolved, there remained a crucial issue that can be put rhetorically as: Can God be a Darwinian? The answer could only appear easy to the two species of fundamentalists that had managed to survive in the Christian or the Darwinian environment. For the fundamentalist Christian the possibility of an evolutionary concept of creation could not be countenanced because it is literally at variance with what he or she found in the Bible. For the fundamentalist Darwinian, who swore by chance mutations, it was preposterous to suggest that the evolutionary tree could have been, in some sense, planned by a deity. For everyone else, the answer was more difficult. To say that the laws of nature imply a lawgiver or to claim that natural selection is simply God's way of ensuring progress left many ill at ease. For what kind of all-wise and all-powerful lawgiver would allow millions of variations to proliferate at random and leave it to the environment to eliminate those that did not happen to fit?

What kind of God would permit the enormous suffering evident in nature? By raising these questions evolutionary theory posed a new challenge to religion. It is not only that the issue became more complex, it is that science seemed to establish that the rise of the human species was wholly dependent on the blind thrust of evolutionary pressures. A theory that banished the realm of mind, value, and responsibility seemed to its critics as much a betrayal of the quest for genuine understanding as was the earlier belief in the invulnerability of scripture. Dogma is as stultifying in science as anywhere else, but unlike the earlier confrontation between science and religion over cosmology, which did not undermine the belief that the world was designed by a benevolent and almighty God, the new debate over the origin of humans seemed to undermine the very notion of Providence. Whether the Sun moves around the Earth or whether it is the other way around does not affect one's interpretation of the basic tenets of Christianity. By freeing believers from looking in the Bible for the science that is not there, the heliocentric theory contributed to focusing attention on the salvation story. It was not clear how the theory of evolution could become an aid in worshipping a God who is personally involved in the history of mankind. But beliefs about the nature and the origin of human beings cannot be settled decisively by observation, experiment, or any other means. An evolutionary process that was essentially subhuman, undirected, and unmotivated would lead to a divorce between man and his roots in nature. The interesting paradox is that so many came to accept God as a Darwinian, albeit a Darwinian with a purpose.

Science, Religion, and the Cartographies of Complexity

David N. Livingstone

There is no surer guide to the territory of science and religion than John Hedley Brooke. He is a superlative cartographer. For several decades he has pioneered the way into this forbidding terrain and mapped its intricate topography with both subtlety and precision. His work has shaped my own endeavors at every scale. On the most trivial level he has furnished me with yet another alliterative term for a lecture aimed at general audiences on the historical relations between science and religion. To *conflict* and *cooperation,* social *competition,* and ideological *continuity* as ways of thinking about the encounter, I have been able to add *complexity* as by far the best way of getting a handle on the issues involved. At a different scale he has infected me with what I would call the Brookean allergy to-*isms.* Brooke's emphasis on the need to disaggregate big concepts and comfortable labels—Baconianism, Newtonianism, Anglicanism, Darwinism, Inductivism, to name but a few—has secured at least one enthusiastic disciple. For these labels conceal as much as they reveal. Like the words *science* and *religion* themselves, such terms are what W. B. Gallie famously called "essentially contested concepts"; they defy specification in terms of transcendent necessary and sufficient conditions.[1]

Brooke is also the master of delicious irony. Time and time again in his many writings, readers are brought face-to-face with the unexpected, with what ought not to have happened, with individuals ending up on the wrong sides in debates. In this role Brooke emerges as an iconoclast in the best sense, demythologizing the icons of science and religion alike. The champions of secular science and religious faith need to be made more aware of the kinds of stories he tells. The enthusiastic coterie of latter-day Darwinians intent on canonizing Charles Darwin and Thomas H. Huxley, for example, need to

From *Historically Speaking* 8 (May/June 2007)

acknowledge the racism and sexism that snake their way through evolution's most sacred texts. Darwin, for example, rejoiced that the Anglo-Saxons would almost certainly exterminate, and replace, the savage races throughout the world and happily called on the support of his eugenically obsessed cousin Francis Galton to insist that if the prudent avoid marriage, whilst the reckless marry, the inferior members tend to supplant the better members of society.[2] For their part latter-day creationists, forever complaining about the evils of social Darwinism, need to be reminded that creationism has been a bulwark of racial prejudice and supremacist ideology. Louis Agassiz, for example, whose name frequently crops up in works of creationist apologetics, was convinced that the different races had each been specially created and displayed a distinctive hierarchy of moral and intellectual excellence. And anti-Darwinian, conservative Presbyterians in the American South had no trouble securing biblical legitimacy for their belief that slavery and racial hierarchy were divinely sanctioned.[3]

Attending to complexities of this sort casts new light on episodes that loom large in the traditional iconography of science-and-religion encounters. The Scopes Monkey Trial in Tennessee in 1925 is a case in point. The formulaic portrayal of this affair as redneck resistance to enlightened science has no doubt served the needs of ideologues of various hues but mostly at the expense of ignoring the fact that the textbook in question, George William Hunter's *Civic Biology: Presented in Problems* (1914), appended to its popularization of evolution declarations insisting that Caucasians were the highest human type and that eugenic practices should be installed to wipe out what he called social parasitism and its cost to society. Given these professions it is hardly surprising that the Great Commoner, William Jennings Bryan, a Democratic candidate for the American presidency, would find the book obnoxious. Comfortable stereotype does not serve us well here at all.

It is precisely the realization that history does not fit preconceived templates that has impelled Brooke to privilege complexity and contradiction over predictability and preconception. And indeed he is entirely right to disavow the suggestion that he is the father of a complexity thesis. For complexity is not to be thought of as a *hypothesis* about the history of science and religion, so much as a *description* of a state of affairs. The more we examine the historical evidence, moreover, the more justification there seems to be for identifying convolution upon complexity. Let me mention one or two further complicating factors.

First, for all the sterling service that historical exercises in demystification have rendered in unraveling the forces at play in science-and-religion disputes directing attention to interests at work in particular clashes, highlighting power plays of various stripes, identifying the significance of rhetorical conventions, excavating hidden or suppressed components of narratives, there is still work to be done in charting the genesis and diffusion of the myths themselves. We have been so intent on debunking fables that we have perhaps ignored another just-as-demanding task: uncovering the archaeology of mythmaking in the history of science and religion and scrutinizing the cultural functions such legends perform. We know a great deal about what really happened in infamous encounters like that between Galileo and the Church in the early decades of the seventeenth century or between Bishop Samuel Wilberforce and Huxley in the nineteenth. Ironically we know much less about how and why the original myth was constructed, the channels through which it circulated, and the ways it was transformed and mobilized in different settings. All this is a way of saying that elucidating the history of myths will help us come to terms with the myths of history. Attending to that agenda would, I think, further serve to confirm the emphasis on complexity that is diagnostic of Brooke's project.

Second, the realization that seemingly simple practices like reading and speaking are themselves polymorphic activities has added further complications to the story of science and religion. Reading never takes place in a vacuum since readers are inescapably located in interpretive communities, which shape their encounters with texts at every turn. As both James A. Secord and Nicolaas Rupke have powerfully shown, scientific texts are differently read in different settings; they are made to mean different things in different spaces. And it thus makes sense to speak of the historical geographies of reading.[4] Because Isaac Newton's, Darwin's, and Albert Einstein's theories were differently construed in different settings, it is mistaken to look for any single response to them from religious believers. Speech, too, is conditioned by space; in different venues different things are speakable and unspeakable. What can be said about a scientific theory and what interlocutors can hear varies from site to site. Talk about science is different in churches and courtrooms, in public spaces and in private salons, in consultations and clinics. Because there are intimate connections between location and locution, students of science and religion need to become more aware of the rhetorical spaces speakers and their listeners occupy rather than taking pronouncements simply at face value.[5] Time and again we find speakers having to control their tongues in particular settings and fitting their rhetoric to the audiences they address.

For all the persuasiveness of Brooke advocacy of complexity, however, his stance has not been without its critics. Here he mentions two charges that have been laid before him. The first attack, namely, that emphasizing complexity has the effect of masking real conflict, stems from a misunderstanding. As

Brooke rightly notes, acknowledging complexity is not intended to deny conflict. To the contrary it is to recognize warfare where it has occurred but to resist elevating it into a ubiquitous metanarrative as some contemporary publicists are only too inclined to do. The second criticism, which revolves around the contention that complexity recedes into particularity and thus disables generalization, is taken more seriously by Brooke. And I, too, sense the force of this accusation. Perhaps one potential way of getting a handle on the problem is to pursue a little more vigorously the cartographic image with which I began these reflections. Just as mapping can be carried out at many scales from the local to the global, so historical encounters between science and religion may be mapped at every scale from the micro to the macro. Some studies will illuminate the relationship at the level of the individual biography or even of a fragmented self. How did individuals deal with the claims of science and religion? Were their views consistent, or did their stance shift over time or space? Others might dwell on the physical and social spaces such individuals occupied and the ways these venues shaped the rendezvous between scientific claims and religious convictions. Still others might focus on the larger communities of which commentators were a part. No doubt such collectives, whether they are institutions, informal associations, or religious traditions, will not be monochrome; but it may well be possible nonetheless to discern certain patterns. As with any distribution map, some generalization must take place at this scale of mapping. It is the same when we move to the regional level. Shifting to this layer of analysis will likely deliver some still more general patterns comparable to, say, an agricultural map or a population map or a median-income map of the state. At every cartographic scale something is lost even as something is gained. As we move from large scale to small scale, specificity is necessarily sacrificed to generality. The key question about any map, therefore, is never simply, Is it accurate? Accuracy is a relative, not an absolute, value. Maps are only more or less accurate with respect to their purposes. A small-scale map of the world is useless for planning a walk around a market town; a large-scale town plan is of no value for determining the employment patterns of a country. It is the same with mapping science and religion: Different scales deliver different insights for different purposes, and it is a mistake to substitute one scale for another.

From my viewpoint, then, Brooke's piece is pleasingly sprinkled with geographical terms: He speaks of sites of knowledge, of mapping local contingencies, of philosophies working according to time and place, of topographies of complexity, of the local and the global. This leads me to suspect that thinking more geographically about the complexities of the history of science and religion might enable us to work more effectively at different scales of analysis but all the while we need to recognize that scale is always relative to purpose.

Notes

1. W. B. Gallie, "Essentially Contested Concepts," *Proceedings of the Aristotelian Society* 56 (1956): 167–98.

2. Charles Darwin, *The Descent of Man* (London: Murray, 1901), 241–42, 945.

3. I have discussed this in David N. Livingstone, *Adam's Ancestors: Race, Religion, and the Politics of Human Origins* (Baltimore, Md.: Johns Hopkins University Press, 2008).

4. See James A. Secord, *Victorian Sensation: The Extraordinary Publication, Reception, and Secret Authorship of Vestiges of the Natural History of Creation* (Chicago: University of Chicago Press, 2001); Nicolaas Rupke, "Geography of Enlightenment: The Critical Reception of Alexander von Humboldt Mexico Work," in *Geography and Enlightenment,* ed. David N. Livingstone and Charles W. J. Withers (Chicago: University of Chicago Press, 1999), 281–94.

5. I have discussed this in David N. Livingstone, "Science, Site, and Speech: Scientific Knowledge and the Spaces of Rhetoric," *History of the Human Sciences* 20 (2007): 71–98.

Response

John Hedley Brooke

I must first thank my respondents for their empathetic comments. To find oneself so generously admitted to the honorary guild of cartographers is a signal honor. It is David Livingstone who has done more than anyone to pursue the map-making metaphors when analyzing the many facets of scientific culture and its modulation in different locations. His essay is a wonderfully succinct account of why location matters and why, as a corollary, an older historiography of science concerned only to chart the history of ideas will no longer do, not even for the development of scientific theories, let alone their cultural implications.

I particularly welcome his recent emphasis on the need to link locution to location. In speaking about the relations between science and religion, there have often been constraints on what it has been politic to say. I have been reminded of a letter I read years ago from Joseph Hooker to Charles Darwin in which the social pressures for conformity were perfectly explicit: "It is all very well for [Alfred Russell] Wallace to wonder at scientific men being afraid of saying what they think. . . . Had he as many kind and good relations as I have, who would be grieved and pained to hear me say what I think, and had he children who would be placed in predicaments most detrimental to children's minds . . . he would not wonder so much."[1] That was as late as 1865 and reminds us again of the ambiguities that could so easily be enshrined in public religious affirmation. Yet, having said that, we are immediately confronted again with the problem of complexity because it does not logically follow that statements designed to appease are necessarily disingenuous. Darwin himself wished to cause minimal offense when writing his *Origin of Species.* But it does seem likely that, at the time of its composition, he really

From *Historically Speaking* 8 (May/June 2007)

did believe that the laws of nature impressed upon matter by the Creator had been deliberately designed.[2]

A heavy insistence on the mapping of local contingencies does, however, run the risk of eliciting unsympathetic reaction, especially if an "anything-goes" postmodernism is suspected. This is not Livingstone's position, but the risk of guilt by association cannot be discounted, as recent comments in the *British Journal for the History of Science* make clear. Reviewing a new textbook John Heilbron complains of the postmodern fetish of the local and contingent.[3] This prompts me, as my rejoinder to Livingstone, to raise the question of whether distinctions may not have to be made between different forms of science and their relative susceptibilities to cartographic analysis. Such analysis is certainly valuable when the issue touches on almost every feature of human identity, as Darwin's science did, and that of *Vestiges* before it. In that case responses, especially the religious, were almost bound to be saturated with cultural preconceptions; the stakes were so high, it would surely be surprising if the philosophizing did not reflect local interests, agendas, and events. But suppose we are discussing harder science and theories, say, concerning the structure of an organic compound. There were, of course, social and professional rivalries involved throughout the nineteenth century as competitive chemists sought to get the right answer first. But we do not so immediately think of the local as deeply affecting the chemical discourse itself and differing responses to it. If we are thinking specifically of the discourse of science and religion, there is the additional consideration that most scientific theories have no bearing whatsoever on theological construction. Do we need to add yet another level of complexity to those defined by the cartographic metaphors? Is not further discrimination required in order to ascertain the different *degrees* to which theories pertaining to *different* sciences repay analysis in the terms Livingstone has indicated?

In William Shea's nicely balanced account of the Copernican and Darwinian issues, there is the suggestion that these two instances of an exacerbated crisis in the Christian churches do have to be differentiated: Whether the Sun moves around the Earth or whether it is the other way around does not affect one's interpretation of the basic tenets of Christianity. The Darwinian theory, by contrast, seemed to undermine the very notion of Providence. This is not an example I would use to press the point for discrimination. After all, as Shea indicates, a moving Earth *was* perceived to have adverse theological consequences in the seventeenth century, and the post-Copernican debates *were* differently structured in different contexts, as I indicated in my essay. The association, typified by Giordano Bruno, of Copernican astronomy with the atomism of the ancients also means that the theory *could* be construed as a challenge to Providence. I do, however, wish to use Shea's distinction in order to make a rather different point and to indicate yet another problem for students of science and religion.

By the nineteenth century biblically based inhibitions and those based on Providentialist concerns were now passé as far as a moving Earth was concerned. Indeed it would be possible to trace the assimilation of Copernicanism through a succession of Christian thinkers, showing how each, in their respective locations, had their reasons for modernizing. The problem is that one could study each thinker in his or her local context, tracing in detail the process of assimilation, and yet still miss something. To focus on a succession of accommodations, each fully contextualized, might miss the cumulative effect of so *much* accommodation. It might miss that moment when the honest doubter suddenly realizes that although accommodation has proved possible for their forebears, the harmonization achieved makes an empty sound.

A classic case in point would be the efforts to harmonize Genesis with geology during the first half of the nineteenth century, before the impact of Darwin was felt. During each decade a new challenge from geology required that a previous harmonization scheme had to be jettisoned. For example the device of postulating a gap in time between the original creation and the events of the first Genesis day had to be superseded by the proposal, first adumbrated by Georges-Louis Leclerc de Buffon, that congruence could be achieved by identifying day with epoch. Such schemes became ever more elaborate, as in the eloquent musings of the Scottish evangelical Hugh Miller, who interpreted the days as the successive days on which the biblical author had inspired visions of the phases the Earth had passed through. But there came a point and it is visible in at least two contributions to *Essays and Reviews* (1860) when the cumulative effect of so much harmonizing was a final admission of dissonance.[4]

The dissonance can then, of course, be exploited by the secular critic of religion. With regard to Shea's example, it is perfectly true that a moving Earth held no terrors for Christian thinkers in the nineteenth century. But a young Thomas Hardy, in his early novel *Two on a Tower,* was nevertheless able to exploit Galileo's telescopic discoveries in a subtle attack on Providence. The issue concerned the purpose of that vast multitude of stars invisible to the naked eye yet identifiable through the telescope. It had been a problem in Galileo's day, but Hardy gave it a new twist. Such stars, evidently, had not been made for us. The bottom line was that therefore *nothing* had. I am not suggesting for a moment that the localizing tendencies of the cartographer are misplaced. They have, for instance, contributed to sharper accounts of both Hugh Miller[5] and Hardy.[6] But I am suggesting that our maps must be sufficiently mobile to allow for both the diachronic and the retrospective restructuring of perceptions. In the seventeenth century to render the Earth a planet was to enhance its significance in the universe. Conceiving of the Earth as a planet promoted it into that superlunary region deemed by Aristotelian philosophy to be immaculate and immutable. It was to escape from what both Galileo and Johannes Kepler considered a dumb place to be—a garbage bin into which everything fell. With reference to the location of Hell, a centric Earth was, after all, diabolocentric. Three centuries later it would be taken for granted, often uncritically, that the displacement of the Earth from the center of the cosmos must have been a relegation: the first of Sigmund Freud's three revolutions (the Copernican, the Darwinian, and the Freudian) that had stripped humanity of its delusions of grandeur.

Notes

1. Joseph Hooker to Charles Darwin, October 6, 1865, in *The Correspondence of Charles Darwin,* ed. Frederick Burkhardt, vol. 13 (Cambridge: Cambridge University Press, 2002), 261–65.

2. Charles Darwin, *The Origin of Species: A Variorum Text,* ed. Morse Peckham (Philadelphia: University of Pennsylvania Press, 1959), 757–59.

3. John Heilbron, review of *Making Modern Science: A Historical Survey,* by Peter J. Bowler and Iwan Rhys Morus, *British Journal for the History of Science* 40 (2007): 118–19.

4. C. W. Goodwin, "On the Mosaic Cosmogony," and Baden Bowell, "On the Study of the Evidences of Christianity," in *Essays and Reviews* (London: Longman, 1860).

5. C. W. Goodwin, "On the Mosaic Cosmogony," and Baden Bowell, "On the Study of the Evidences of Christianity," *Essays and Reviews* (London: Longman, 1860).

6. Claire Tomalin, *Thomas Hardy, the Time-torn Man* (New York: Viking, 2006), 189–97.

PART 3

Herbert Butterfield and the Scientific Revolution

Reassessing the Butterfield Thesis

Peter Harrison

Like so many familiar historical categories—the Middle Ages, the Renaissance, the Reformation, the Enlightenment—the Scientific Revolution has undergone something of an identity crisis in recent years. For historians of science both terms in the expression *Scientific Revolution* have come to be regarded as problematic.[1] Revolution is said to be misleading because the relevant transitions took place over a rather more protracted time period than the term would normally warrant. A. Rupert Hall's *Scientific Revolution: 1500–1800,* to offer an instructive example, implies a "revolution" of some three hundred years' duration. While it is possible to contract the chronological scope of the putative revolution to the more manageable one hundred years or so between Galileo and Newton, this still leaves us with a rather less climactic event than such analogous political upheavals as the French or Russian revolutions. Neither is it easy to identify specific events or occasions that might act as markers that signal the commencement or, for that matter, the completion of the Scientific Revolution. There is no equivalent, in other words, of the fall of Rome or the storming of the Bastille. Some have thought that the publication of Nicolaus Copernicus's *De revolutionibus* in 1543 might play such a role, and the coincidental appearance of the *De humani corporis fabrica* of Andreas Vesalius in the same year might seem to add weight to the status of this date as the logical *terminus a quo* for this particular revolution. Yet when we consider its immediate impact, the only thing revolutionary about Copernicus's work was its title. *De revolutionibus* aroused little controversy at the time and for almost half a century attracted few converts. Robert Westman has convincingly argued that there were at most ten genuine Copernicans in Europe before the year 1600, considerably fewer, I suspect,

From *Historically Speaking* 8 (September/October 2006)

than those currently living in the United States who are committed to the geocentric system.[2] It took the combined efforts of Johannes Kepler, Galileo, René Descartes, Isaac Newton, and others to establish the credentials of the heliocentric hypothesis, and even at the end of the seventeenth century, there were many who remained unconvinced. In sum we can legitimately ask whether the term *revolution* is the right word here and, beyond issues of semantics, question the idea of a radical discontinuity between medieval and modern approaches to the study of nature.

The term *scientific* is, if anything, more problematic. Strictly speaking *science* in its modern form did not appear until the nineteenth century. Indeed some have spoken of a second scientific revolution that took place at that time. The study of nature in the sixteenth and seventeenth centuries was conducted in a variety of disciplines—natural philosophy, natural history, the mixed mathematical sciences (such as astronomy and mechanics), anatomy, astrology, and alchemy. According to the traditional divisions of the sciences inherited from the Middle Ages, natural philosophy concerned itself with the causes of change and motion in the world but typically excluded mathematical treatments. These took place in the mixed mathematical sciences, which dealt with such artificial human constructions as machines or with the idealized motions of celestial objects. The mathematical accounts of the movements and positions of heavenly bodies that characterized astronomy were largely kept separate from the physical and causal discussions of the same phenomena that formed the subject matter of natural philosophy. Natural history was a historical enterprise in the sense that it generally restricted itself to the descriptions of living things. Moreover, until well into the seventeenth century, it was primarily a textual activity, for knowledge of animals, plants, and minerals was then gained primarily by studying the authoritative texts of ancient writers or the more recent digests of those ancient works. That aspect of natural history that dealt with plants was closely allied with medicine ("physick"), for the vast majority of cures were herbal. Knowledge of astrology was also considered to be important for physicians, for celestial influences played a significant role in determining the health of the individual. The general point here is that while these enterprises may seem to have direct analogues in the subdisciplines of the modern sciences, natural philosophy looks like physics, for example, and natural history like biology—in terms of their methods and their relationships to each other, these earlier approaches to nature differed in quite significant ways from what we would call science. To exploit the political analogy once again, had there not been a recognizable constitutional entity "France" in the eighteenth century, there could not have been a *French* revolution. Because there was no unitary entity "science" in the seventeenth century, so the argument goes, there could have been no *scientific* revolution.[3]

None of this amounts to a decisive argument against the usefulness of the idea of a scientific revolution, however. Recognition that this revolution had an extended duration was actually one of the features of Herbert Butterfield's book, yet this realization did not prompt him to dispense with the term. Neither was he entirely oblivious to the significance of the different disciplinary arrangements of the time. It must be said, however, that he lacked the perspective provided by recent scholarship on the nature of the early modern disciplines. A concerted focus on such matters has been a major historiographical concern of early modern historians of science over the past two decades. The question remains, then, whether our new sensitivity to these kinds of issues necessarily rules out the idea of *scientific* revolution.

In my view even if we take seriously the nature of the disciplinary divisions of the early modern period and allow that, strictly speaking, there was then no science, it may still be possible to recast changes in approaches to the natural world in less-anachronistic language while preserving the view that an important historical shift took place during this period. We might speak, for example, of the transformation of natural philosophy or of the tendency for natural philosophy to subsume within itself other approaches to the natural world. Consider the standard case of the introduction of mathematics into natural philosophy. As we have seen, in the Aristotelian taxonomy of the sciences, mathematics was supposed to be kept distinct from natural philosophy. According to Aristotle mathematics dealt with human abstractions rather than with nature itself. A number of figures traditionally associated with the Scientific Revolution challenged this embargo and imported mathematical methods into natural philosophy. Kepler thus insisted against the traditionalists that mathematics had a legitimate place in accounts of the true physical arrangement of the heavens. Newton expressed a similar view in the very title of his masterwork, *Philosophiae naturalis principia mathematica* (The Mathematical Principles of Natural Philosophy).[4] There is something quite revolutionary about these developments.

There were other fruitful attempts to transgress traditional disciplinary boundaries during this period. Francis Bacon articulated an ambitious plan to link natural history and natural philosophy, thus combining historical and philosophical approaches to nature. The former would provide an atheoretical assortment of facts upon which the causal explanations of natural philosophy were to be constructed. The flaw of previous natural philosophers, in

Bacon's view, was that they had prematurely offered explanatory claims (the philosophical task) before sufficient factual evidence had been accumulated (the historical task).[5] William Harvey's discovery of the circulation of the blood can also be thought of as resulting from a willingness to ignore received disciplinary demarcations. According to Andrew French, Harvey's originality lay in introducing into the sphere of medical anatomy questions that were essentially those of the natural philosopher.[6] The former discipline had been directed to the practice of medicine, rather than the contemplation of natural philosophical questions. It rested on the authority of Galen, rather than Aristotle. It was by posing the philosophical question of what it was to be a heart that Harvey came to his radical conclusion. What all of this suggests is that even if we are not dealing here with a revolution within some long-standing intellectual activity, "science," we can nonetheless point to important changes that took place within natural philosophy and other disciplines concerned with the natural world. These changes led to new ways of investigating nature, some of which have been preserved in the modern sciences.

The attitudes of key figures involved in the transformation of these disciplines are also revealing. When reading the works of those typically identified as major contributors to the Scientific Revolution, the modern reader cannot help but be struck by the sense that at least some imagined themselves to be living in an era of revolutionary change in the realm of human knowledge. Across these apparently disparate disciplines, thinkers considered themselves to be participants in something that was quite novel. In the sphere of natural history, navigators and explorers reported that they had discovered many new things and secrets that had been unknown to the ancients.[7] Kepler's *Astronomia nova* (1609) proudly announced itself as the herald of a *new* astronomy. Bacon's *Novum organum* (1620) was so entitled in order to draw attention to the fact that it was to replace the *old* organon, a grouping of Aristotle logical works that occupied a central place in the university curriculum. Galileo spoke of two new sciences, while Descartes emerged from his stove-heated room convinced that he would found a marvelous *new* science based on mathematical principles.[8] Excitement about the prospects of new knowledge assumed apocalyptic proportions in seventeenth-century England, where, owing to Francis Bacon's invocation of the eschatological imagery of the book of Daniel, the advancement of learning was interpreted as a prelude to the end of the present age. All of this is suggestive of self-conscious attempts to revolutionize knowledge. If the relevant historical actors viewed their own activities in this light, the idea of a scientific revolution would seem to be more than just a scholarly construction foisted onto a range of quite disparate developments in the early modern period. The changes taking place within a number of different spheres of learning, moreover, would seem to be more than unconnected coincidences.

What would lend further weight to the idea of revolutionary change is the identification of plausible causal factors that might have precipitated changes across these apparently heterogeneous areas of interest and produced the kinds of attitudes evident among some of the main protagonists. Historians have become justifiably wary of specifying unitary causes for significant historical transitions such as these. Unlike scientific data historical events are typically overdetermined. However this need not deter us from attempting to identify the kinds of factors that may have inspired these thinkers with the idea that they were destined to be the progenitors of new forms of knowledge. One possible trigger has been mentioned already. The deficiencies of ancient knowledge in the sphere of natural history became painfully obvious with the voyages of discovery. A growing awareness of the limitations of the inheritance of classical learning contributed both to a diminution in the authority of the ancients and to a turn away from the works of human authors to the book of nature itself. Beyond this, however, it is worth reconsidering the two historical episodes that Butterfield mentions in connection with the Scientific Revolution but relegates to the status of mere internal displacements within the history of medieval Christendom: the Renaissance and the Reformation.

The Renaissance, assuming for the moment that there was such a thing, witnessed the revival of ancient Platonism, Epicureanism, and Skepticism. Together these were to challenge the monopoly exercised by Aristotelianism and provide alternative philosophical resources for natural philosophy. Platonism thus elevated the status of mathematics and arguably played a part in what has traditionally been called the mathematization of nature. Interest in Epicureanism led to a revival of ancient atomism, which informed much of the new matter theory of the early modern period. A growing familiarity with the arguments of the ancient skeptics promoted challenges to those who espoused dogmatic certainties, and these were often identified as Aristotelians. The philological endeavors of humanist scholars also played an important role in exposing gaps in the transmissions of ancient knowledge and in the cumulative errors of generations of copyists, particularly in the sphere of natural history. Once again the perceived unreliability of the written corpus of knowledge transmitted from antiquity prompted investigators to the firsthand study of nature itself. So while it can be admitted that the category *renaissance* suffers to a degree from the same kind of credibility problem as the Scientific Revolution, we can still make the modest claim that the renewed interest in classical authors characteristic of the period paradoxically contributed to a growing confidence in the capacity of moderns to surpass the ancients in the sphere of learning.

Arguably the Reformation was even more important than the Renaissance in terms of its role in the emergence of the new sciences of nature. For one thing it is the most secure historiographical category of the three. To be sure it is now fashionable to speak of plural reformations: Protestant, Catholic, and Radical reformations; or German, French, and English reformations; or even princely, peasant, and urban reformations. That notwithstanding no historian, as far as I know, has declared that there was no such thing as the Reformation. It must be understood, of course, that the processes of religious reformation included early modern Catholicism. Neither should it be forgotten that in an important sense Protestantism was initially a Catholic phenomenon insofar as the first generations of Reformers were Catholic. We need also to remind ourselves that the Reformation had far-reaching consequences for the whole of the West that had only an indirect relation to the theological ideas of the major reformers. Bearing this in mind it is possible to identify a number of ways in which religious reformation had an impact on new developments in the sciences of nature.

Protestant Reformers, and Luther in particular, were sharply critical of the pagan Aristotle and of the baleful influence of peripatetic philosophy on the university curriculum and theology in particular. Misgivings about the omniscience of Aristotle reinforced the reservations expressed by some Renaissance thinkers and gave license to those seeking alternative ways of pursuing natural philosophy and natural history. The full scientific implications of a Christian doctrine of creation were also explored for the first time, for while Aristotle's assertion of the eternity of the world had never been accepted in the Christian Middle Ages, the ramifications of his view had never been fully purged from natural philosophy. Taking seriously the idea of the world as a divine artifact meant the end of Aristotle's distinction between the sciences of the natural and of the artificial and thus undermined one important justification for keeping mechanics and physics separate. If God was a mathematically inclined artisan, moreover, this raised the possibility, endorsed in different ways by Kepler and Descartes, that God had instantiated real mathematical relations in the natural world. As the examples of both Galileo and Descartes illustrate, one did not have to be a Protestant to take advantage of Aristotle's increasing vulnerability on points such as these. (By the same token the biographies of Galileo and Descartes also suggest that the expression of views contrary to Aristotelian orthodoxy was often easier for those living in Protestant domains.) The loosening of Aristotle's grip on the Western mind, which Butterfield tends to attribute to the Scientific Revolution, had thus begun before Kepler, Galileo, and Newton announced their novel views.

Also linked to the Reformation was the revival of a relatively pessimistic Augustinian anthropology that stressed, among other things, the fallenness of the human intellect and the imperfections of bodily senses. This tendency was most conspicuous in Calvinism but was manifested also in some forms of Lutheranism and in Jansenism. Those who subscribed to this anthropology found themselves unable to accept an optimistic and uncritical Aristotelianism that presumed that the world was more or less as it presented itself to the senses. As Butterfield himself put it, Aristotle's physics was "based on the ordinary observation of the data available to common sense."[9] To Calvinist critics, this common sense had been corrupted by sin.

Incapable of penetrating to the essences of things and constitutionally precluded from arriving at the demonstrative certainty that for Aristotle was the hallmark of genuine *scientia,* sinful human beings had to be satisfied with probabilistic knowledge. Such knowledge, moreover, was acquired only after long periods of arduous observing, experimenting, and collecting. Recourse was had to artificial instruments in order to rectify imperfect senses, and the unique incapacities and biases of the individual were to be overcome by practicing philosophy in a communal setting. Experimental natural philosophy, particularly as conducted in England, was thus underwritten by a theological anthropology promoted by the Protestant Reformers. In short if the new mathematical natural philosophy gained ground at the expense of Aristotle's ideas about mathematics, experimental natural philosophy was grounded in a view of human nature that was incompatible with Aristotle's sanguine and uncritical epistemology.

A good case can also be made that some Protestant ideas and practices contributed to the desacralization of nature. Curious as it may seem, the Protestant insistence on the primacy of biblical authority and the superiority of literal readings of scripture played a significant role here. Throughout the Middle Ages allegorical readings of scripture had been commonplace. These entailed a particular view of the natural world, for allegorical readings assumed that natural objects bore theological and moral meanings. Thomas Aquinas, following Augustine, had explained that while the literal sense lay in the meanings of particular *words,* the allegorical sense lay in the meanings of the *objects* to which the words referred.[10] Allegory thus provided a means of transporting the reader from the meanings of the words of scripture to the manifold symbolic meanings of the objects of nature. The natural world, in this scheme of things, was a rich repository of transcendental truths. In asserting the primacy of the literal sense, the reformers were in effect insisting that while words could have referential functions, natural objects could not. The reformers elevation of scriptural authority summarized in Luther's motto *sola scriptura* further eroded the theological significance of the natural world by diminishing nature's role as a source of divine truth. Divested of its theological meanings, the natural world now lay open to alternative, scientific ordering principles. Further impetus for the disenchantment of nature came from Protestant doubts about the veracity of contemporary miracle claims from their iconoclasm and from their radical contraction of the number of sacramental channels through which divine grace could find its way into the world.

The idea of connecting the Reformation and the rise of science, it must be said, is a rather risky proposition, and most previous attempts have met with grief. Best known, perhaps, are the claims of Robert K. Merton, who correlated scientific activity with Puritanism in seventeenth-century England. Merton's definition of Puritanism was too loose, his definition of science too essentialist, and his apparent indifference to French and Italian achievements lessened the generality of his thesis.[11] Still there remains something compelling about a boldly stated historical thesis of a kind that is becoming a rarity today, and Merton's idea, borrowed from Max Weber, that connects Calvinist "this-worldly asceticism" with some features of Western modernity still has something to commend it. But apart from the claims of more recent historians, it is also telling that some leading early modern natural philosophers connected these two events, suggesting that religious reformation had provided a precedent for a more general reformation of learning. This sentiment, not surprisingly, was particularly strong among English Protestants. Francis Bacon asserted that the reformation of the Christian church was but the first stage of a more comprehensive renovation of all learning. This historicist commitment was widespread in seventeenth-century England.[12] Neither was it uncommon in other parts of Europe to associate innovators in the sciences of nature with leading figures in the religious-reform movements. Copernicus and Paracelsus, according to one writer, were the Martin Luther and John Calvin of natural philosophy. Kepler was known as the Luther of astronomy.[13] So it was that some of the traditional heroes of the Scientific Revolution drew both inspiration and social legitimacy from the almost contemporary upheavals in the realm of religion. For them, too, it seems that the religious reformation was the primary event and, as Bacon expressed it, the spring of subsequent reformations.

In sum, then, Butterfield articulates the view that there was a scientific revolution, albeit one of extended duration, and that this particular revolution was largely responsible for giving Western modernity its characteristic features. Of possible competing events—the Reformation, the Renaissance, and the advent of Christianity—only the last had a comparable influence. What I have suggested is that the Scientific Revolution is still a useful category, provided that we hedge it with appropriate qualifications. So, too, are the Reformation and Renaissance. In the contentious business of ranking these in terms of their influence, however, I part company with Butterfield. The Reformation—and here I admittedly take a broader conception of religious reformation than that assumed by Butterfield—was a major factor in creating the kind of world in which a particular kind of natural philosophy could take root and flourish. And insofar as it made this contribution, it is to be placed before the rise of science both chronologically and in terms of its impact on the West. In any case the religious upheavals of the sixteenth century, apart from any impact on the emergence of science, saw the confessionalization of Europe, contributed to the birth of the modern state, and initiated the processes of secularization. (It is tempting to throw capitalism into this mix, but at this point I imagine that my respondents already have enough controversial claims to reckon with.) Butterfield was right to argue for the importance of the Scientific Revolution (although we might quibble with the exact terminology), but he went too far in claiming it to have been largely independent of and more influential than the Reformation. That said his *Origins of Modern Science* has weathered the test of time better than most works of a similar vintage. It is remarkably well informed, most of its factual claims would still stand today, and even major contentions of the kind presently under discussion are carefully nuanced. It is a still a book from which we can learn.

Notes

1. For summaries of the relevant issues, see Steven Shapin, *The Scientific Revolution* (Chicago: University of Chicago Press, 1996), 1–4; David C. Lindberg, "Conceptions of the Scientific Revolution from Bacon to Butterfield," in *Reappraisals of the Scientific Revolution,* ed. Lindberg and Robert Westman (Cambridge: Cambridge University Press, 1990), 1–26; and A. Rupert Hall, "Retrospection on the Scientific Revolution," in *Renaissance and Revolution: Humanists, Scholars, Craftsmen and Natural Philosophers in Early Modern Europe,* ed. J. V. Field and Frank James (Cambridge: Cambridge University Press, 1993), 239–50.

2. Robert Westman, "Copernicus and the Churches," in *God and Nature: Historical Essays on the Encounter between Christianity and Science,* ed. David C. Lindberg and Ronald L. Numbers (Berkeley: University of California Press, 1986), 76–113.

Cf. Herbert Butterfield, *The Origins of Modern Science,* 1949 (New York: Free Press, 1997), 44.

3. Some would suggest that the same is true of contemporary science, which is said to comprise quite heterogeneous activities that share no common method.

4. Johannes Kepler, *Mysterium Cosmographicum,* 1621, trans. A. M. Duncan (Norwalk, Conn.: Abarus, 1999), 123; Andrew Cunningham, "How the *Principia* Got Its Name: Or, Taking Natural Philosophy Seriously," *History of Science* 28 (1991): 377–92.

5. Francis Bacon, *Novum Organum,* in *The Works of Francis Bacon,* ed. James Spedding, Robert Ellis, and Douglas Heath, 14 vols. (London: Longman, 1857–74), 4:71; Bacon, *Great Instauration,* in *Works,* 4:28–29; and Bacon, *De Augmentis,* in *Works,* 4:295, 299, 362.

6. Andrew French, *William Harvey Natural Philosophy* (Cambridge: Cambridge University Press, 1994).

7. See, for example, Nicholas Monardes, *Joyfull News out of the Newe Founde Worlde* (London: Willyam Norton, 1577), fol. 34v.

8. Galileo, *Dialogues Concerning Two New Sciences* (1638; New York: Macmillan, 1914); René Descartes, *Oeuvres de Descartes,* 13 vols., ed. Charles Adam and Paul Tannery (Paris: Cerf, 1897–1913), 10:156–57, 179.

9. Butterfield, *Origins,* 16.

10. Augustine, *De doctrina christiana* 1.2 2; Thomas Aquinas, *Summa theologiae* 1a.1, 10.

11. For a brief but searching critique of Merton, see the review of his *Science, Technology and Society in Seventeenth-Century England* by one of my respondents, William Shea, in *Philosophy of Science* 41 (1974): 89–90.

12. Francis Bacon, *The Advancement of Learning,* ed. Arthur Johnston (Oxford: Clarendon, 1974), 42. See also Thomas Sprat, *History of the Royal Society* (London, 1666; Whitefish, Mont.: Kessinger, 2003), 371 (in the Kessinger edition); Thomas Culpeper, *Morall Discourses and Essayes* (London: Adams, 1655), 63; Samuel Hartlib, Hartlib Papers 48 17, Sheffield University Library, reproduced in *The Great Instauration: Science, Medicine, and Reform, 1626–1660,* ed. Charles Webster (London: Duckworth, 1975), 524–28; and Noah Biggs, *Mataetechnia Medicinae Praxeos. The Vanity of the Craft of Physick . . . : to the Parliament of England* (London: Calvert, 1651).

13. R[ichard]. B[ostocke]., *The Difference between the auncient Phisicke . . . and the Latter Phisicke* (London: Walley, 1585), sigs. cviii.v, hvii.v; Charles Webster, *From Paracelsus to Newton* (Cambridge: Cambridge University Press, 1982), 4.

Response to Harrison

William R. Shea

The Scientific Revolution, meaning roughly the period that runs from Copernicus to Newton, is now so deeply entrenched in the literature that it is hard to believe that it was only given broad currency in Herbert Butterfield's *Origins of Modern Science* in 1949. Whereas nineteenth-century historians claimed that the great changes that catapulted Europe into the modern age were the Reformation and the Renaissance, Butterfield saw the major breakthrough in the twin advance of scientific conceptualization and factual discovery that began in the sixteenth century. I agree with Peter Harrison that Butterfield captured a major aspect of the historical shift that took place at this time, and I will stress in this response some of the reasons why his thesis still holds.

We need only reread the famous aphorisms at the beginning of Bacon's *Novum organum* to be reminded that our way of viewing the world changed in the seventeenth century:

> I. Man, the minister and interpreter of nature, understands and does as much as he is able to observe of nature through reason or his senses. He neither knows nor is capable of more.
>
> II. Neither the unassisted hand nor the understanding left to itself is worth much. Results are obtained with the help of instruments, which are required by the understanding as well as by the hand. And just as instruments provide the hand with motion or regulate it, so they stimulate or safeguard the action of the intellect.
>
> III. Science and human power are one and the same.

From *Historically Speaking* 8 (September/October 2006)

The shift is clear: Knowledge has become power to be used not to contemplate nature but to improve it. Such a change would have been impossible without tools that opened new vistas. The most spectacular was the telescope that revealed a new heaven. The alleged difference between the Earth and celestial bodies crumbled with the sight of mountains and crevasses on the moon, and geocentrism was challenged by the discovery that Venus revolved around the Sun and not the Earth. Ptolemy was literally disarmed by the telescope, and the countless stars that Galileo witnessed ushered in a new awesome feeling about man's place in the world that John Donne captured:

> The Sun is lost and the Earth, and no man's wit
> Can well direct him where to looke for it.[1]

Unlike the human eye, telescopes can be improved, and as they became better, things were seen more clearly and described more rigorously. Lenses could bring the heavens closer; they could also delve inside objects. The microscope, which was used by the Linceans in Rome to study flowers and insects, slowly acquired the scientific importance with which we are familiar. But the infinitely small proved as much of a challenge as the infinitely large, and man's position in the cosmos became increasingly problematic. The awesome character of the human predicament is captured in one of Pascal's *Pensées* where he describes the most minute of insects, the mite, which reveals under magnification a world of even more minute parts that can be divided over and over again until the last atom is reached. But why stop here? This atom itself could conceal "an infinity of universes, each with its firmament, its planets, its earth, in the same proportion in which we will find again the same proportion as in the first."[2]

The telescope and the microscope thus gave new scope to speculation on the nature of the infinite, thereby preparing the mental space required for the new calculus and the role of infinitesimals. On a more mundane level the thermometer and the barometer brought temperature and weather within the purview of quantification. By the middle of the seventeenth century Descartes and Pascal's brother-in-law Florin Périer were exchanging barometric observations with colleagues in Sweden. A few years later the Accademia del Cimento in Florence was designing beautiful and practical instruments that had a medical as well as a purely scientific interest.

Instruments to facilitate computation were also a hallmark of the Scientific Revolution. While in Padua, Galileo devised a general-purpose mechanical calculator or "compass" capable of solving any practical mathematical problem (multiplication, division, sighting, triangulation, calibration, and so on) swiftly, simply, without previous mathematical education. With a compass of his own invention, Descartes showed how to solve two famous mathematical problems of antiquity: the duplication of the cube and the trisection of an angle. Galileo's discovery of the isochronisms of the pendulum found medical application in the pulsilogium of Santorio Santorio and, with a crucial improvement introduced by Christiaan Huygens, provided the foundation of the modern clock. The seventeenth-century conquest of time and the introduction of clockwork precision and regularity in all walks and manifestation of life are now so obvious and so commonplace that we forget their momentous impacts on civilization. We may feel a sense of liberation to be without a watch for a short period of time, but imagine the crisis and the chaos that would instantly paralyze every developed country of the world if all the clocks were to stop or go on strike.

The new facts that swam into the consciousness of the seventeenth century were not all uncovered with the aid of instruments. New flora and fauna were brought back from voyages of discovery to other continents, and anthropological reports about the "noble savage" proved unsettling. But all this new information would not have been disseminated so rapidly and so effectively without the technological breakthrough represented by letterpress printing from movable type. Book printing was a boon for learning in all fields, and illustrations enabled communication of technological innovation by other means than words alone. What could only with great difficulty be conveyed in sentences now leapt to the eyes from eloquent drawings in such books as those of the sixteenth-century engineers Georg Agricola and Agostino Ramelli. What made Andreas Vesalius's masterpiece *De humani corporis fabrica* so striking were the dramatic and artistic illustrations that raised structural anatomy to a new level. Simple diagrams rendered possible what would have daunted the most skilled of writers. Galileo heaved a sigh of relief when he could rely on a sketch to illustrate how a piece of machinery worked: "I imagined a device that I can explain much better with a sketch than with words."[3] Such possibilities were not available to an earlier age.

The new instruments, however, would have had a limited impact if they had not been coupled with deep innovations in the realm of ideas and mentalities. In astronomy, for instance, the gradual acceptance of the idea that the apparently stationary Earth is really dashing around the Sun had a number of important and unexpected results. At the general level of philosophical discourse, it gave a new urgency to the need to distinguish between the subjectivity of sense perception and the objectivity of scientific language, and it renewed interest in the hypothetical or the realist status of scientific theories. At a practical level Harvey was probably led to discover the circulation of the blood by likening the heart, the king and ruler of the microcosm, to the Sun, the source of warmth and life in the macrocosm.

The new science of motion, which began with Galileo's study of falling bodies and projectile motion and culminated in Newton's law of universal gravitation, changed the way physics was understood and ushered in a new relationship between art and nature whereby unnatural conditions (for example, dropping objects in a vacuum) began to disclose the hidden truth behind natural operations. The idea that man can only know what he can make surfaced in scientific discussions and was actively propagated by Marin Mersenne. What man can know about himself was profoundly modified by the discovery that the Earth had existed for a much longer time than had hitherto been believed. New theories about the formation of the Earth's crust initiated a difficult but relentless reevaluation of man's place in nature.

The Aristotelian network of ideas that was dismantled during the seventeenth century was more than a scientific hypothesis. It defined the purpose of science, the means of knowledge, and the relations among man, nature, and God. The common belief in the fundamental stability of the existing natural structures carried with it a conviction of their wise design. When Descartes attempted to explain human physiology in mechanical terms, he subverted the belief that mankind could understand the final causes or purposes for which things had been made. The mechanical philosopher sought the design or scheme of things in the original ordering of matter and its laws. The God of History became the Supreme Artisan. Robert Boyle, who wholeheartedly accepted the Cartesian separation of mind and matter, saw in the clock, and particularly in the great cathedral clock of Strasbourg, the greatest work of ingenuity contrived by man: It symbolized the highest known intelligence and exemplified in the most elaborate detail the adaptation of means to ends. The image of clockwork led Boyle in the direction of a designer instead of the view that we are mere cogs in a machine. The Scientific Revolution turned God into a watchmaker but not one that was blind.

Bacon and Descartes may have flirted with the arcane but they went on to marry reason. The Scientific Revolution marked the demise, however gradual, of the hermetic concept of knowledge. An open science is essential to an open society, and Butterfield saw that this was a main thrust of the scientific movement. Books on magic, alchemy, and astrology still circulated in the seventeenth century, but they were increasingly read in a context where obscurity ceased to pass muster for depth and where pretentious claims were subjected to experimental confirmation.

In the Middle Ages philosophy had flourished in the cloisters as the handmaid of theology. In the seventeenth century natural philosophy found a new home and a new atmosphere in scientific institutions. The academies in Rome, Florence, London, Paris, and Berlin saw the necessity of corporate investigation and international cooperation. They offered a forum for discussing scientific problems and methods. They also played a vital role in transmitting information and encouraging the spread of science through specialized journals. If they added prestige to the role of the scientist, they also stimulated interest among a large number of amateurs and thus created a scientific public.

The goal of a unified science and the desire of achieving mastery over nature had been present in the alchemical tradition. That the same ideals should be shared by the Scientific Revolution is, of course, understandable, but it should not make us forget that the way of achieving those ends was profoundly different. The new cast of mind favored experimental control and public debate whereas hermeticism dreamed of leaping over rationality itself. What Butterfield underestimated was the role that Christianity played in the emergence of modern science, and this has now been set right by Harrison and other scholars.[4] The Scientific Revolution was not without ancestry, but it ushered in a radically new way of understanding the world and ourselves.

Notes

1. John Donne, "An Anatomy of the World. The First Anniversary," *Complete Poetry and Selected Prose,* ed. John Hayward (London: Nonesuch, 1972), 202.

2. Blaise Pascal, *Oeuvres Complètes,* ed. Louis Lafuma (Paris: Editions du Seuil, 1963), 526.

3. Galileo Galilei, *Opere,* ed. Antonio Favaro, 20 vols. (Florence: Barbèra, 1890–1909), 8:62.

4. Peter Harrison, *The Bible, Protestantism, and the Rise of Natural Science* (Cambridge: Cambridge University Press, 1998). See also Kenneth J. Howell, *God's Two Books: Copernican Cosmology and Biblical Interpretation* (South Bend, Ind.: University of Notre Dame Press, 2002) and, from a different vantage point, Rodney Stark, *The Victory of Reason: How Christianity Led to Freedom, Capitalism, and Western Success* (New York: Random, 2005).

The Butterfield Thesis and the Scientific Revolution

Comments on Peter Harrison

David C. Lindberg

I admire Peter Harrison's willingness to march boldly into this particular jungle. The question of the "Scientific Revolution" (Was there such a thing? If so, when did it occur, and what were its defining characteristics?) has become a favorite pastime of historians of early modern science who thrive on frustration and conflict. Forty years ago we knew what the Scientific Revolution was and had a pretty good idea when it occurred and what its causes were. But unanimity is now a distant memory.

I would like to begin where Harrison ends, with that most difficult of questions about the Scientific Revolution: its possible relationship to theological developments associated with the Protestant Reformation. I have read most of Harrison's publications on questions of science and religion, including *The Bible, Protestantism, and the Rise of Natural Science* (1998) and the manuscript of *The Fall of Man and the Foundations of Science, 1500–1700* (2007), and I believe that he has established himself as a leader in the field of early modern science and religion. We get a taste of his views in his gentle dismissal of the Merton thesis, which attempted to explain seventeenth-century English empiricism as the offspring of the Puritan ethos. I think it is just as well that he has chosen not to discuss a more recent explanatory attempt to find a religious cause for seventeenth-century empiricism—namely, the influence of voluntarist theology—a subject that Harrison has dealt with elsewhere.[1]

From *Historically Speaking* 8 (September/October 2006)

Harrison defends two religious influences on the practice of seventeenth-century science. He argues that the emphasis within Reformation Protestantism on the literal sense of scripture led to "erosion" of the allegorical meanings that had been attached to natural objects during the Middle Ages and a shift toward literal readings of nature. The result was a transformation of natural history. The other religious influence acknowledged by Harrison in this essay straddles the divide between Protestants and Catholics. This is "pessimistic Augustinian anthropology," which cast doubt on the full recovery of the intellectual powers lost with the fall of Adam, leaving the human race no alternative but to compensate for that loss by diligent observation and experimentation.[2]

I do not share Harrison's concern with the semantics of the expression *Scientific Revolution.* It is true that we have no universally accepted definition of either *science* or *revolution.* But the same is true of most interesting words. We have no universally accepted definition of *Middle Ages, Renaissance, fall of Rome, art, music, religion, philosophy, historian, Christian,* and so on. All of these are abstractions with debatable meanings that vary from one linguistic community to another and from person to person within a given linguistic community. They are labels that we require if we are to communicate with one another, and by definition they are fuzzy. With practice we gain the ability to discern from context what is meant on a given occasion. To quibble about the label *Scientific Revolution* is thus (in my opinion) a waste of time; we should save our quibbling for the occasions for which quibbling was invented—namely, discussion of such things as scientific beliefs and practices, rather than what to name them.

On one point I must disagree with Harrison. In his search for innovative seventeenth-century developments that dramatically influenced the way "science" (substitute the word of your choice) was practiced, Harrison claims that natural philosophy (what most of us now call "physics") and mathematics lived separate lives until the seventeenth century, when they were finally united in matrimony. This is a widely held belief among historians of science, going back to Pierre Duhem at the beginning of the twentieth century and restored to respectability by Robert S. Westman in a 1980 article.[3] But it is demonstrably false, as anybody who studies the sources closely will discover.

I am aware that the previous sentence evinces a level of dogmatism rarely encountered in reputable historians—at least in print. But here the evidence is irrefutable. In more than forty years of research in the primary sources of ancient, medieval, and Renaissance optics (both Islamic and European), I have not yet come across any major optical scientist (*perspectivist* was the medieval Latin term) who was not concerned with *both* the mathematics *and* the physics (or physiology) of light, color, and vision.[4] My investment in the astronomical sources is of far shorter duration, but I have immersed myself for the past six months in that literature, and I can make the same claim. From Ptolemy through medieval Islam and medieval Europe to the Renaissance, the universal goal—not necessarily the achievement but the goal—was to develop a *physical model* that would yield accurate *numerical predictions* of planetary behavior. In short the notion that there were uncrossable disciplinary boundaries between physics and mathematics is an illusion based on unfamiliarity with the relevant texts. To be sure Aristotle differentiated between mathematics and physics. But he transgressed the boundary himself, and we have no evidence and no reason to believe that medieval and Renaissance followers took Aristotle's differentiation as even a mild prohibition. If I have begun to sound like a crusader on this issue, you read me correctly. It is time for historians of science to give up the illusion of a premodern separation between the mathematical and the physical.

I conclude with a caveat and an assumption that calls for critique. The caveat concerns the relationship between, on the one hand, the global change typically envisioned (I believe) by defenders of the seventeenth-century Scientific Revolution and, on the other, the individual disciplines that made up the world of actual scientific practice. Disciplines have different histories, proceed at different rates, and answer to different methodological canons. Peek beneath the surface of a global scientific revolution, and you will find that it rests on an array of disciplinary revolutions. It follows that we need to narrow our focus from the macroscopic to the microscopic if we hope to write an enduring account of the seventeenth-century Scientific Revolution.

The assumption that I question (I have no reason to believe that Harrison is guilty of holding it) is the ubiquitous association of revolution and discontinuity. The Scientific Revolution, it is generally assumed, entailed repudiation of the medieval scientific tradition. I believe, however, that attention to developments at the *disciplinary level* will reveal that the revolution was built on a medieval foundation with the aid of resources derived from the ancient and medieval past. To construct his planetary models, for example, Copernicus employed geometrical devices transmitted from their ancient Greek and medieval Islamic creators by the classical tradition. The foundational first proposition of Galileo's treatise on the science of motion is the "Merton rule," borrowed from mathematicians associated with Merton College, Oxford, in the fourteenth century. And Kepler created his revolutionary theory of the retinal image exclusively with the geometrical resources of medieval *perspectiva.* Many additional examples could be produced. Nothing in these remarks is meant to diminish or minimize the creative achievements of seventeenth-century scientists but to identify the origin of some (but by no means all) of the resources creatively employed by those scientists to revolutionize the disciplines within which they labored.

Notes

1. Peter Harrison, "Voluntarism and Early Modern Science," *History of Science* 40 (2002): 63–89. For an opposing view, see Margaret J. Osler, *Divine Will and the Mechanical Philosophy: Gassendi and Descartes on Contingency and Necessity in the Created World* (Cambridge: Cambridge University Press, 1994). The central claim of theological voluntarism is that God's freedom of action is without any limitation, from which it follows that our rational capacities will inevitably fail in any quest to determine the inner workings of the created world in which we live. Consequently our only hope of acquiring such knowledge is to invest heavily in observation and experimentation. To clinch this plausible argument, its defenders claim to find a close correlation among seventeenth-century scientists between the holding of voluntarist theology and the practice of observation and experiment—a correlation that I find dubious.

2. See Harrison's *Fall of Man and the Foundations of Science, 1500–1700* (Cambridge: Cambridge University Press, 2007) for a deluge of persuasive evidence.

3. Pierre Duhem, *To Save the Phenomena: An Essay on the Idea of Physical Theory from Plato to Galileo,* trans. Edmund Doland and Chaninah Maschler (1908; Chicago: University of Chicago Press, 1969). Robert S. Westman, "The Astronomer's Role in the Sixteenth Century: A Preliminary Study," *History of Science* 18 (1980): 105–47.

4. For a succinct history of ancient and medieval optics, see David C. Lindberg, *Theories of Vision from al-Kindi to Kepler* (Chicago: University of Chicago Press, 1976). For a deeper look, see Lindberg, *Roger Bacon's Philosophy of Nature* (Oxford: Clarendon, 1983).

Response to Harrison

Butterfield's *Origins of Modern Science* and the Scientific Revolution

Charles C. Gillispie

For those of us attempting to inaugurate teaching of the history of science after the war, Herbert Butterfield's *Origins of Modern Science, 1300–1800* was literally a godsend. He began the cultivation of a largely untilled field like a deus ex machina bestriding, albeit unpretentiously, the discipline of history proper. Not only did he set the example of how to write a narrative history of technical material but he also wrote a book we could give undergraduates to read. And this may have been even more valuable. The current generation can scarcely imagine the conceptual and stylistic poverty of what passed for the literature half a century ago. Apart from Arthur O. Lovejoy's *Great Chain of Being* (1936), E. A. Burtt's *Metaphysical Foundations of Modern Science* (rev. ed. 1932), the chapters on English, French, and German scientific styles in J. T. Merz's *History of European Thought in the Nineteenth Century* (1896–1914), and James B. Conant's *On Understanding Science, an Historical Approach* (1947), none of which was properly historical, the pickings were thin and thorny. That said it is ironic that Butterfield is best known to historians at large for his devastating and perhaps overdone critique *The Whig Interpretation of History* (1931), for there is no more classic an example of Whiggishness in the historiography of science than his *Origins of Modern Science.* The notion of a delayed scientific revolution in chemistry is instance enough.

It is no reflection on Butterfield to say that he popularized, or better publicized, an analysis developed by Burtt and more deeply by Alexandre Koyré

From *Historically Speaking* 8 (September/October 2006)

in *Études galiléennes* (1939) and *From the Closed World to the Infinite Universe* (1957). Thanks to them—though not new with them—the very concept of a scientific revolution was of crucial importance to our thinking. It gave us a framework with which to contrast our accounts of ancient and medieval science—the word may be thought anachronistic, but *scientia* does mean knowledge—and within which to develop the story line leading to modernity.

To rehearse only the highest lights: Copernicus's *De revolutionibus orbium coelestium* (1543) gives Kepler the basis for mathematically formulating and observationally confirming the laws of planetary motion (*Astronomia nova* [1609], *De harmonica mundi* [1619]); experimentation as discovery, full fledged in William Gilbert (*De magnete,* [1600]), is joined to experimentation as confirmation of mathematical formulation by Galileo in founding classical kinematics if not quite dynamics (*Discorsi . . . à due Nuove Scienze* [1638]); Galileo's trial and conviction for the *Dialogo dei Massimi Systemi del mondo* (1632) is science as theater; what with the Royal Society, the Académie Royale des Sciences, Stevin, Beekman, Huygens, Paracelsus, Hooke, Mariotte, Boyle, et al., the stage soon grows crowded; Vesalius's *De humani corporis fabrica* (1543) appears coincidentally with Copernicus's *De revolutionibus.* The background of Renaissance artistic realism thus naturalized sciences of life, wherein developments parallel those in the physical sciences. Dissection and experiment at the hands of Girolamo Fabrici, Realdo Colombo, and others culminates in Harvey's experimental discovery of the circulation of the blood (*De motu cordis* [1628]), confirmed by quantitative reasoning. Except for an increasingly rigorous botany, the life sciences waited until the early nineteenth century for biology to emerge, on the one hand, from medical concerns, mainly physiology and anatomy, and, on the other hand, from natural history, mainly classification of the animal kingdom.

Throughout the seventeenth-century protagonists insist on the novelty of their work. The adjective *new* appears in many titles. Moderns rebel against ancients in the battle of the books. It is always the business of philosophers, not so much to innovate themselves as to see deeply into what is happening to knowledge. Bacon does so for the relation of experiment to the new knowledge of nature, Descartes for the relation of mathematics to the new knowledge of nature, and Spinoza for the relation of the new knowledge of nature to theology and religion. Finally in *Principia mathematica naturalis philosophiae* (1787), Newton unites laws of celestial motion with laws of terrestrial motion—Galileo with Kepler—and proceeds to the experimental foundation of the science of colors in the *Opticks* (1704). The curtain thereupon comes down on the opening act of the history of modern science, a convergent plot so far, and goes up without an intermission on the divergent pattern of successive acts wherein the plot is united only by method and consequences.

Engaged in other things I have not kept up with all the revisionist writings on the reality of the Scientific Revolution. My sense is that like other historical complexes, the Renaissance, the Enlightenment, and the Industrial Revolution, it will survive attacks on its cogency. The critique is a function of the shift in focus that has transpired in the historiography of science since the 1970s. The word *science* has two senses. It connotes, on the one hand, a body of knowledge bearing on the structure and forces of nature. It connotes, on the other hand, the activities for obtaining and applying that knowledge. The interest of many historians of science has shifted in the last generation from the content to the practice of science in the past, often the very recent past. Until late in the eighteenth century natural philosophers and naturalists had more in common with their predecessors in the ways they lived and worked than they did with scientists of the nineteenth and twentieth centuries. The notion of a second scientific revolution turns on what institutionalized the change, namely, professionalization and discipline formation. I have written about that among other things in *Science and Polity in France: The End of the Old Regime* (1980, 2004) and *Science and Polity in France: The Revolutionary and Napoleonic Years* (2004).

With respect to science as knowledge, knowledge of the world in which we live, the situation is different. It is, of course, true that scholarship has traced certain technical antecedents in statics, kinematics, mechanics, optics, chemistry, and methodology to medieval and Islamic origins. It is also true that vestiges lingered. Kepler cast horoscopes, but Newton did not. Newton steeped himself in alchemy, but the next generation did not (nor were those the researches that made him the principal founder of classical physics). That a break occurs in a splintering manner and not cleanly does not make it incomplete. For chemistry is not alchemy, astronomy is not astrology, momentum is not impetus. There is a categorical difference between Aristotelian motion as actualization of a potential and inertial motion as uncaused and more widely between Aristotelian and modern conceptions of change in general. Nothing in medieval or Arabic science modified the world picture of living on a mother Earth at the center of a finite spherical cosmos. None of it touched the belief in correspondences between heaven and Earth, between the stars and human affairs, personal, political, and military. None of it threatened the medieval Christian belief in a world created for us by God and governed by His Providence, wherein humanity fulfills its destiny. Apart from the revival of Epicureanism in the Renaissance, there is no carryover from the ancient cosmos to the infinite and meaningless universe, totally indifferent to us, in which modern humanity lives, has its being, and seeks (perhaps vainly) for sparks of life beyond a solar system whose eschatology is a function of the thermodynamics of the Sun. Those implications have still to be widely accepted and perhaps never will be, but that does not conjure them away. Nor were they perceived by natural philosophers themselves. Kepler and Newton were devout and godly men, though one has one's doubts about Galileo. I have mentioned Spinoza. He saw the point early on (*Tractatus theologico-politicus* [1670] and *Ethics* [1678]). So, too, as Matthew Stewart has persuasively suggested, did Leibniz, who, however, sought to evade it in his *Theodicy* (*The Courtier and the Heretic: Leibniz, Spinoza, and the Fate of God in the Modern World* [Norton, 2006]). Many will disagree, and everyone has a right to an opinion, but it seems to me impossible to deny that the secure framework of a cosmos centered on humanity was gone. That there were differences of opinion about the relation of spiritual to physical and material reality is indicative enough. To the best of my knowledge, which is always subject to correction, there were none such in the Middle Ages. At the very least the issue of the relation of science to religion began with the Scientific Revolution, tragically in Galileo's case, philosophically in Spinoza's.

It is of no significance for the change in sensibility that Newton's infinite Euclidean space, unchanging in time, has given way to the relativistic evolving universe of the Big Bang. That is a function of the development of the sciences themselves. In much if not all of their content they, too, have moved far beyond the sciences of the seventeenth century. But cognitively if not socially speaking, they are enterprises of the same sort. Thomas Kuhn was my career-long friend and colleague. But I do not fully go along with the argument of his brilliant *Structure of Scientific Revolutions* (1962). In his terms, and I do agree with this one, the content of medieval and modern science and the successive world pictures are incommensurable the former with the latter.

Whether or not one wishes to designate what happened in science between 1543 and 1687 as the Scientific Revolution or the first phase in the history of modern science is a merely semantic question. In either case there is no way to measure whether its effects on the history of Western civilization were greater or less than those of the Reformation, ably set forth in themselves and in their intermingling with science by Peter Harrison. When the question is expanded to incorporate world history, however, there can be little doubt about it. I have suggested elsewhere that the concept of the Scientific Revolution should be enlarged to encompass both Renaissance behavior patterns and the voyages of discovery (*The Edge of Objectivity*, preface to 2nd ed. [1990]). The activities of a Brunelleschi, a Leonardo, a Michelangelo, a Prince Henry of Portugal, a Vasco da Gama, a Christopher Columbus, an Amerigo Vespucci, a Ferdinando Magellan, a Francis Drake, all were animated by the forward-looking outlook that later formed a Galileo. Theirs was the instinct that knowledge finds its purpose in action and action its reason in knowledge; that if a problem can be solved, it should be solved; that if a constructive thing can be done, it should be done. Travels whether by sea or land in antiquity or later in Asia and the Far East entailed legend, adventure, or commerce. What differentiates the European voyages from those and relates them to science is that they were involved with, though perhaps not motivated by, the problem of how the world is made—in a word, with knowledge. Not for nothing were they characterized as voyages of "discovery."

Those voyages were, moreover, the bearers of that civilization that created modern science and that in consequence came to dominate the world until yesterday, for whatever mixture of good and ill. Together with the technology that attends and empowers it, science, both as knowledge and activity, has been the only aspect of Western civilization that others have fully embraced. What the imperial court in China welcomed from Jesuit missionaries was calendrical astronomy, not Christianity. Beginning with the Japanese in modern times, Asian polities have deliberately adopted and assimilated science, if only in self-defense at the outset. And, no matter what the contrasts in cultural background, they practice science as well as we do and no differently. It is, finally, ironic to reflect that the decline of Islam began with its gradual abandonment of the scientific enterprise in which it had outshone Europe throughout the Middle Ages.

Although these last considerations were not what Butterfield had in mind, I do think they justify his sense that the creation of modern science—the signal word in his title is *Origins*, not *Revolution*—has had the most far-reaching effects of any set of events since the decline and fall of the Roman Empire and the Christianization of Europe.

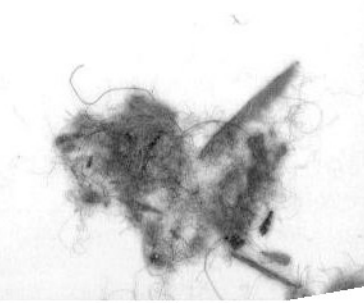

Rejoinder to Gillispie, Lindberg, and Shea

Peter Harrison

Let me first thank my respondents for their thought-provoking contributions. I am pleased to be in such illustrious company. I am also pleased that we seem to be in general agreement about the value of Butterfield's contribution and the importance of retaining the category "the Scientific Revolution," albeit with some caveats.

William Shea provides a useful reminder of the importance for this revolution of such things as scientific instruments and the printing press. Indeed Shea offers a gentle corrective to my somewhat-intellectualized account of the changes that took place in the sciences of this period. Whereas I pointed largely to conceptual revolutions and the reasons for them, he has rightly shown that material factors played a pivotal role in the production of the new forms of knowledge. Shea also highlights the importance of a new emphasis on the mastery of nature, to which I shall return at the end of this rejoinder.

There are several other matters raised in the responses on which we might engage in profitable discussion. However I shall restrict myself to two of the more important, raised in the main by David Lindberg and Charles Gillispie. These are, first, the issue of the putative separation of natural philosophy and mathematics in medieval science and, second, the question of whether concern with the early modern conception of *science* is largely a matter of semantics.

Lindberg disagrees with my contention that natural philosophy and mathematics operated as two distinct disciplines until the seventeenth century. This criticism is pertinent to our present discussion insofar as it relates to an increasingly common view that what was distinctive (or "revolutionary") about early modern physics was its mathematical character. This issue

From *Historically Speaking* 8 (September/October 2006)

has been the subject of an amicable disagreement between myself and Lindberg that extends beyond this forum. I must say that I am increasingly inclined to believe that Lindberg has a point and that at the very least his objections warrant further investigation. My own view is consistent with a body of influential secondary literature and also relies on the formal classifications of the sciences that we encounter in the medieval and Renaissance texts. These do seem to support an important distinction between mathematical sciences and natural philosophy, with only the latter understood as providing true causal accounts of natural phenomena. This distinction also explains the special status of the so-called mixed mathematical sciences. Yet whether the relevant practitioners were themselves inclined to observe these distinctions is another matter entirely, and it may be that in practice these formal distinctions were simply ignored.

After all it is not clear that practicing scientists of any era have conformed to the theoretical accounts of scientific activity provided by philosophers of science. Certainly I have no grounds for disputing Lindberg's claim that those medieval thinkers concerned with optics dealt with both mathematical and physical aspects of the relevant phenomena. That said there does seem to have been a long-standing convention according to which certain astronomical models had value by virtue of the fact that they saved the phenomena while not necessarily being true accounts of physical reality. This can perhaps be traced to Thomas Aquinas's attempt to square Ptolemy's mathematical astronomy and its eccentrics and epicycles with Aristotle's natural philosophy and its homocentric spheres. Aquinas declared that the former was not necessarily true. Although it was consistent with appearances, the phenomena could be saved in some other way not yet understood. This stance seems to be invoked again in Osiander's infamous preface to Copernicus's *De revolutionibus,* where the heliocentric hypothesis is described as "a calculus that fits the observations" although again "not necessarily true." The principle appears again in Galileo's equally infamous (and disingenuous) disclaimer at the end of the *Dialogue concerning the Two Chief World Systems.*[1] It is also suggestive, as I noted in my original article, that Kepler seems to have believed that natural philosophers would be opposed to his use of mathematics in physical astronomy.[2]

One possible direction for a resolution of this disagreement lies in the important difference between optics and astronomy. It may be that the so-called instrumentalism of mathematical astronomy was a function of the lack of direct sensory access to the relevant phenomena. This would be consistent with Lindberg's suggestion that the *goal* of astronomy, albeit unrealized, was a true physical model that provided accurate numerical predictions. Peter Barker and Bernard Goldstein, who have also challenged the common view of Renaissance astronomy, have suggested that the putative instrumentalism of astronomy was a consequence not of any disciplinary commitment but of the want of direct knowledge of the physical arrangement of the heavens. Sixteenth-century astronomers, they conclude, should not be regarded as fictionalists but as "perpetually frustrated realists."[3]

These are questions that clearly require further investigation. If Lindberg is correct, then a number of early modernists will have to relinquish their cherished conviction that seventeenth-century natural philosophy, on account of its mathematical character, is distinctively different from that which came before. As it relates to the present discussion, this would weaken the case for a strong discontinuity between medieval and early modern natural philosophy.

The other major issue concerns the identity of *science* (as opposed to *natural philosophy* and so on) and whether such a concern is mostly to do with mere semantics. Lindberg and Gillispie both imply that in taking up the issue of the identity of early modern "science" I have yielded to an unhealthy and misplaced preoccupation with semantics. While I agree that it is important not to quibble about the meaning of words if nothing hangs on the outcome, I do believe that attending to the way in which historical actors used such terms as *science* and *natural philosophy* has important payoffs for the historian. Thus, for example, my concern with how *science* and *scientific* were used arises in part out of a sensitivity to precisely the kind of issue to which Lindberg helpfully draws our attention in his concluding remarks—namely, that the early modern disciplines that we regard as "scientific" have discrete histories and methods, and that when we carefully scrutinize the "scientific revolution" we encounter "an array of disciplinary revolutions." One reason that I believe we need to be careful when using the term *science* as an umbrella for these distinct activities is that such caution helps us avoid conferring upon them a unity that they do not have.

Gillispie's response also shows why it is important to think carefully about the meanings of *science*. This term, he suggests, has two senses—one referring to a body of knowledge, the other concerned with the activities through which such knowledge is gained. Gillispie rightly observes that in recent years historians have turned their attention from content to practice. But careful consideration of the contemporary meanings of *scientia* enables us to extend our historical investigations even further. In the Middle Ages and early modern period, *scientia* did indeed refer to knowledge in general. In a more specific and technical sense, it meant certain or demonstrable knowledge in keeping with the Aristotelian ideal of scientific knowledge. But there was also a third sense, one that gradually began to disappear during the period under discussion. This was the idea that *scientia* was a mental habit, literally an "intellectual virtue" possessed by the investigator.[4]

Like the moral virtues *scientia* was a personal attribute that could be developed with practice. Throughout the Middle Ages and into the Renaissance it was believed that "scientific" activity—more correctly, the practice of natural philosophy—would cultivate this inner quality of *scientia.* This was consistent with a view of the nature of philosophy itself, which held that being a philosopher was primarily about contemplation and becoming a certain kind of person.[5] So if we ask what was philosophical about natural philosophy, an important part of the answer will involve reference to contemplation and cultivation of the intellectual virtues.

This brings us, in conclusion, to one of Shea's key points: With the Scientific Revolution "knowledge has become power to be used not to contemplate nature, but to improve it." Francis Bacon epitomizes this turn away from the contemplation of nature to its mastery. Part of what marks this change, I would suggest, is a shift in the meaning of *scientia* and a corresponding reconceptualizing of the philosophical orientation of natural philosophy. This shift was promoted in part by Ockhamist and Protestant critiques of Aristotelian notions of virtue and by an insistence by some within the Reformed Confession that science was not a mental habit but a body of doctrine.[6] From this time onward, natural philosophy began to separate itself from philosophy proper, and to that extent it started to look much more like modern science.

Tracing the changing meanings of such terms as *science* and *natural philosophy* thus provides crucial insights into such major historical transitions as the Scientific Revolution.

Notes

1. C. H. Lohr, "The Medieval Interpretation of Aristotle," in *The Cambridge History of Later Medieval Philosophy,* ed. Norman Kretzmann, Anthony Kenny, and Jan Pinborg (Cambridge: Cambridge University Press, 1988), 80–98; Nicolaus Copernicus, *On the Revolution of Heavenly Spheres,* trans. C. Wallis (New York: Prometheus, 1995), 3; and Galileo, *Dialogue concerning the Two Chief World Systems,* trans. Stillman Drake (New York: Modern Library, 2001), 538.

2. See also the discussion of the views of Michael Maestlin, Kepler's teacher, in Charlotte Methuen, *Kepler's Tübingen: Stimulus to a Theological Mathematics* (Aldershot, Hampshire, U.K.: Ashgate, 1998), chap. 5.

3. Peter Barker and Bernard R. Goldstein, "Realism and Instrumentalism in Sixteenth-Century Astronomy: A Reappraisal," *Perspectives on Science* 6 (1998): 232–58.

4. Aristotle, *Nicomachean Ethics,* 1103a, 1143b, 1149; Thomas Aquinas, *Summa theologiae* 1a.82, 3; 1a2ae.52, 2; 1a2ae.57, 1; and *Disputed Questions on Truth* 2, Q. 17, A. 1.

5. See, for example, C. Condren, I. Hunter, and S. Gaukroger, eds., *The Philosopher in Early Modern Europe: The Nature of a Contested Identity* (Cambridge: Cambridge University Press, 2006); Pierre Hadot, *Philosophy as a Way of Life* (Oxford: Blackwell, 1995); and Stephen Gaukroger, *Francis Bacon and the Transformation of Early-Modern Philosophy* (Cambridge: Cambridge University Press, 2001), 5.

6. C. H. Lohr, "Metaphysics," in *The Cambridge History of Renaissance Philosophy,* ed. Charles B. Schmitt and Quentin Skinner (Cambridge: Cambridge University Press, 1988), 535–638, especially 632f.

PART 4

Progress in History?

Progress in History

Bruce Mazlish

The subject of "progress in history" is one that branches out in many directions. To start with it, is helpful to divide the topic into two parts. On one side we need to talk about progress in *history,* in the sense of disciplinary advancement. On another side the main part of the discussion to follow, we need to talk about progress *in* human history, that is, *results,* contested though they are, of this advance as shown according to the progress of the discipline itself. In other words we can talk about progress, that is, advance in the conditions of humanity, only in the light of the discipline of history's achievement of greater mastery of both the empirical and theoretical aspects of the subject.

History as a discipline has become highly sophisticated in regard to the use of archival materials and their interpretation in the light of theories from other fields, especially the social and natural sciences. All sorts of subfields, ranging from micro to big history, from cultural to multicultural history, from environmental to global history, are flourishing. They enable us to pose the question of progress in history, in the second sense that I have posited, in a more sophisticated and complex manner than hitherto.[1]

Before going further, however, I want to look more closely at the *idea* of progress. Needless to say I will be giving a brief and highly selective account, but it can help set the stage for my argument. *Progress* is a key word in modern history. Here it has taken on the quality of myth. Its history emphasizes the confluence of factors that entered into its becoming a/the dominant idea of the past held in the West. The seventeenth century is the seedbed. Then, especially in England and its American colonies, religion in the form of millenarianism foresaw progress on Earth in a thousand-year utopia, before the return of Christ and the day of final judgment.[2] In conjunction with the

From *Historically Speaking* 7 (May/June 2006)

Baconians and their belief in the advancement of knowledge, with science as cumulative and pursued by many hands and minds (eventually, for example, in the Royal Society), these two movements underlay the battle of the ancients and the moderns and the triumph of the latter around 1698. Add to this the political revolutions, first of 1640 and then of 1688 in England, and the idea of progress came to reign supreme for the next few centuries. One of the achievements of modern historiography is a more acute knowledge of how religion and science interacted in early modern history. Until the eighteenth century, even in the light of the Galilean episode, they can be seen as cooperating in a joint task: pursuing the track of God in the natural world. That pursuit showed "progress" occurring in regard to scientific knowledge. In the eighteenth century the secular aspect of advancement became more prominent, and in the nineteenth century this took the form of evolutionary biology, that is, Darwinism. It is at this time that the division of religion and science emerged in its most pronounced form. It was in this form that it survived until the twentieth century.

Then faith in progress came under great strain. The terrible breakdown in civilization (another key word, with much affinity to progress) during World War I, the devastations of World War II (and its resultant dissolution of Western colonialism), the experience of the Holocaust, and the advent of nuclear war and its continuing threat all called into question the idea of progress. When Michel Foucault, for example, attacked the progressive notion that humanity was moving toward freedom, he did so by arguing that, instead, we were moving toward a carceral society. Though this is obviously an inversion of the Whig belief in progress as freedom, it symbolizes the way in which postmodernism has helped make the idea of progress fall out of favor today.

With the idea of progress discredited (though in fact I believe wrongly so in the sweeping way it is often done), ideas of contingency and randomness came strongly to the fore.[3] But they were not to have things all their own way; a contest arose between them and such ideas as directionality and even a revived teleology. What is striking is to see how the same problems arose in both the natural and the human sciences. Indeed it is also striking to see the efforts to break down the division between them and weave a seamless explanation of both the natural and the human past. Thus the nineteenth-century embrace of a division between *Geisteswissenschaft* and *Naturwissenschaft* on epistemological and other grounds has been replaced by a desire to unite them (as, for example, in big history). In fact this is a return to an earlier map of knowledge in which *historia* was not so much a discipline as a means of approaching material in any field.[4] The rapprochement is manifest not only in history and philosophy per se but also in all areas upon which they touch. This is especially true in regard to religion. Because Christianity is a historical religion, based on an account of its coming into being, the fight over the establishment of cordial relations between religion and other branches of knowledge is intense.

We have already noted the role of religion in the origins of the idea of progress. The next thing to look at is the role religion has played in the progress of the human species, a secular inquiry. Then we would have to inquire into the question whether, philosophically rather than historically, religion and the idea of progress are basically antithetical elements, in the sense that religion's ultimate aims are not of this world. We must ask whether Christian and non-Christian efforts to achieve salvation are most often an obstacle to more material forms of progress. To raise such questions is to plunge us into the thickets of, among other things, Weberian notions of the fostering nature of Calvinism in the coming of capitalism and so forth.

Obviously the questions and answers are multifold. As such they illustrate the increasing complexity of life and our ways of studying it. Findings will not stay within their original territory. To take merely one example: In biology, we have the controversy between, say, Stephen Jay Gould and Simon Conway Morris. Gould argued, on the basis of work with the Burgess Shale fossils, that evolution was and is a matter of contingency. There was no necessity in the appearance of intelligent life on Earth. In earlier eons the accidental survival of a particular form led to life as we know it. If that form had not survived and left its sequel, Gould would not be arguing his case. Conway has argued in contrast that evolution has been constrained to follow certain paths, with the result being intelligent beings. He, too, is an authority on the Burgess Shale but has come to a diametrically opposite conclusion from Gould. While both men believe in Darwin and natural selection—Conway is careful to disassociate himself from creationism—one emphasizes contingency while the other stresses directionality.

As we know such esoteric arguments in biology have entered into contemporary political/religious debates. Creationism cum intelligent design has become a partisan matter, seemingly remote from the work of biological science itself. Thus Pope Benedict declares: "We are not some casual and meaningless product of evolution. Each of us is the result of a thought of God."[5] His words are partly echoed by the Dalai Lama, who has written a book on the convergence of science and spirituality. School boards in the United States have gotten into the act and pontificated on the equal merits of the mere "theory" of Darwin's evolutionary biology and that of intelligent design. President George W. Bush has advocated equal time for both theories. The same debates over contingency and necessity take place in the social science fields, often drawing inspiration from what is going on in the religion/science debates.

I want now to stand back a bit from the debates and offer some of my own observations. It is first necessary to recognize that there is enormous confusion over terms. This is matched by a debilitating confusion of the levels of discourse, that is, shifting ideas, for example, from biology to the social sciences without acknowledging their variations in meaning. To say that contingency rules in the natural world, as shown by Darwinian evolution via natural selection, is not the same as to assert that it rules in history. The valences are different. Thus when we ask what historians should take from the natural sciences, we must be extremely careful not to assume a one-to-one translation. As I shall argue at more length later, we can find progress in history at the same time as contingency reigns there.

Let me test this assertion in a few specific areas. Economic "science" is a relatively late development, necessarily dependent on developments in economic life itself, that is, the blooming of market relations around the seventeenth and eighteenth centuries in concurrence with a so-called commercial revolution, succeeded by an industrial one. In an attempt to impose order on the chaos of multiple transactions and exchanges that characterized actual life, Adam Smith and others postulated a providential oversight that led individuals pursuing their own self-interest to foster increased production. This was read as progress, that is, material progress. It was also progress in knowledge, especially as Providence was replaced by the invisible hand, and the latter seen as working through the laws of supply and demand.

Was this also progress in the spiritual realm? Were people better off morally as well as spiritually? Or to put it in present-day terms, did economic development necessarily mean greater well-being? Clearly it is necessary to disaggregate the notion of progress and ask in what areas, at what price, and at what stage of history. It seems to me indubitable that progress has occurred in the sense of increased production of goods and services, in the complexity of economic life, and in many parts of the world a raised standard of living. But some argue that there has been retrogression in the quality of life, in the degradation of the environment, and perhaps in human relations as well.

There is nothing new in the points I am raising: They are the staples of contemporary discussion in regard to economic life. The answers clearly depend on the level at which the questions are being raised and at what point in the cycle of development we are raising them. More important for our discussion we may be able to argue that progress of a variegated kind has been occurring in economic life at the same time as nonprogressive losses and that both are a matter of contingency. At which point we must throw in the notion of human agency. Are all these developments not a matter of necessity, of powerful currents running in a given direction irrespective of human desires beyond the immediate one of material betterment? Or can humans stick an oar in the current and steer the frail bark of humanity toward a more desirable end?

Though not systematically I have tried to suggest the extraordinarily complicated and confused way in which issues such as contingency, agency, directionality, and even teleology interact and leave us devoid of simple conclusions. It is all very well for biologists and their religious opponents to mutter on about contingency versus necessity, about random variations versus intelligent design, and all this may be important but of little relevance to the pedestrian world of historical phenomenology. At the shorter-term, lower level of explanation, the same issues must be addressed in terms that are proper for the historian and not the biologist. The same words, for example, *contingency* and *necessity*, have different meanings in the two spheres. If this be disunity, let us make the most of it!

Let me address the same concerns in one other area: the political. This is probably the most chaotic field, the least amenable to a scientific approach in spite of the misnomer, political science. The clash of personalities, parties, and particular forms of self-interest is constant, operating in seemingly kaleidoscopic fashion. To the historian groping a way through this shifting landscape, the one constant appears to be contingency. It can be expressed as the want of a nail syndrome. As Machiavelli reminds us (speaking of contingency as fortune), Cesare Borgia's illness of a few days at a crucial moment deprived him of his usual energies and left him unable to cope with the situation confronting him, with momentous local consequences. Closer to our own times, 9/11 can be seen as an accident—it did not have to happen—shaping everything henceforth around and beyond it. So, too, there was nothing necessary about the invasion of Iraq; it was the choice of the Bush administration and hence contingent. Yet its consequences are of enormous significance, some of which can be analyzed.

How in the face of such statements can I then claim room for any directionality in history? The answer lies in a resort to the idea of levels and long-term developments. For example for the Bush administration and its supporters, a shift to increased freedom in the world is evident. For more serious thinkers, like Francis Fukuyama, this is tied to the "end of history." Of course such assertions return us to the debate about freedom: What is it, is it all of one kind, and so forth? Bypassing this debate I would like to seize hold of one small piece of the argument, that concerning the extension and spread of human rights. It seems to me that the historical record supports the idea that deep and sustained currents have been swirling toward the expansion of human rights.[6] This is not to suggest a linear expansion. Indeed there have been numerous setbacks and elisions. Nevertheless human rights have been "progressing."

Their expansion is closely connected to that of present-day globalization. Defined in minimal fashion as increasing interdependence and interconnection of human beings, one can discern forms of globalizing in centuries before the present—highlights may be seen in the Silk Road of the seventh century, the explorations of the fifteenth century, and so on—with an extraordinary spurt after World War II. The fact that the very term *globalization* only becomes prevalent sometime in the 1960s or 1970s is symbolic of the explosion that had taken place in the second half of the twentieth century. Was this development necessary? Was it inevitable? Predetermined? My answer would be, "Of course not." Nor is it teleological; it could falter and even cease tomorrow with massive nuclear blasts or an onrush of global warming. Still while thus recognizing that globalization is contingent, one can also argue that it represents a direction in history, which in turn rests on powerful forces and currents that make it a likely outcome.

At this point, having dealt in scandalously brief fashion with a few historical problems in regard to economics and political science—mere hints at the complex issues involved in terms of contingency and directionality—I want to return to the biological and religious fields touched on earlier. Here, along with others, I would assert that the debate over evolutionary theory is, in fact, a political one, part of the culture wars taking place in the United States. Intelligent design was disposed of by Darwin in his original work. It is no longer a serious scientific question. It is, however, a serious religious question. Much of religion arises from the human need for meaning in life—and death. Indeed much of humanity cannot face the thought of its own death. Belief in an afterlife saves us from this awful fate. Such belief is intimately tied to a belief in a God, and this by extension to the notion of intelligent design. The universe is designed for *us*. We are not insignificant specks in an infinite, evolving universe but the center of creation.

Such assertions are beyond the reach of scientific method, the core of science. Clearly I am making categorical statements. In this same spirit I would assert that religious truth is in a separate category from scientific truth; therefore it can have nothing of substance to say in the latter domain. And, as has been frequently asserted, natural science has little substantive to say about religious concerns. Science does, of course, embody a cast of mind that throws doubt on many assertions of religion (which, needless to say, is not monolithic but has many heavenly mansions).

It is not natural science but social science—and for present purposes I am presenting them as disconnected—that mainly bears on our knowledge of religion.[7] True a question such as whether religious beliefs confer a survival advantage on their holders trenches on biological science and may be subject to the inquiries of a particular scientific method. But mainly it is the work of Émile Durkheim and fellow sociologists, the comparative researches of anthropologists, and the studies of historians (what, incidentally, ever happened to the Higher Criticism?) that is pertinent. And it is in these latter fields that questions of complexity, contingency, necessity, directionality, human agency, and unintended consequences emerge in terms that can be given specificity.

Where does all the above leave the benighted historian? First of all I suggest that he/she remember that the present debate about Darwinian evolution and intelligent design, the context at the moment for the large questions discussed above, is overwhelmingly an American one. This conclusion accords with the fact that the United States stands out as a religious country in the developed world. Survey after survey shows this fact, whether in terms of belief in the devil, the afterlife, the need for religion in daily life, and the like. Although echoed in the Catholic Church and elsewhere, the intelligent-design debate itself is a parochial one. However the debates about contingency and directionality in history are not. They are unavoidable contentions for the historian who lifts his/her head from the archival materials even for a moment. Which is why the topic "progress in history" is of critical importance.

A few summary statements are in order. The first is that evolutionary theory, Darwinian in inspiration, is the essential starting context, the background, for the historian trying to understand the human past. This is not, I have argued, the same as importing specific biological ideas into history. The drive for unity of knowledge is powerful and seductive. It can, as I have suggested, mislead. It can divert the historian (cum social scientist) from understanding humanity and its past by specifically historical means.

The fact is that evolutionary theory tells us that about thirty-five thousand to forty thousand years ago, with the appearance of *Homo sapiens* (and the dates and empirical evidence supporting them are a constant subject of debate and new evidence), the human species took a turn to *cultural* evolution. A creature emerged endowed with symbolic abilities, the capacity to think in terms of past, present, and future, and a gradual realization of its own mortality. Other animals can be said also to have consciousness, but none is self-conscious to the same degree as is the human animal, and certainly none has a historical consciousness. Indeed these are all late developments in the case of *Homo sapiens* (with times and details still debatable).[8]

Something new, therefore, has been added to previous evolution. A new domain has emerged. It requires its own decisions as to how and when contingency, directionality, and the like operate. Contingency, especially in the form of unintended consequences, is clearly present, thus constraining human agency to unpredictable shapes. So, too, however, is directionality. As Shakespeare realized, "There is a divinity that shapes our ends, rough hew them as we may." However, as we realize, divinity is an unclear force, perhaps a mere construct of the human mind. Substitutes have been suggested in the form of an invisible hand, historical materialism, and most recently chaos theory.

The ability to discern patterns is a trait characteristic of many animals. It may be that humans alone can discern patterns over extended time. One such pattern may be progress. But of what sort and shape and emerging by what means are not at all clear. Amid the complexity of history, the historian can only grope toward answers. Some observers see the species as taking on godlike features in this effort. Such a view appears to entail the human as God, the self-realization of His aspirations. I would like to conclude with the more modest aspiration of the human as historian who seeks in cultural evolution the empirical evidence, embedded as it is in theory, that either supports or denies directionality. As Darwin said in his *Origin of Species,* "There is grandeur [enough]" in such a view.

Notes

1. As is evident, I am bypassing questions of objectivity, reality, narrative, metanarrative and so forth that would occupy a philosopher of history. They do not affect the point I am making here.

2. A classic account, though focused on America, is Ernst Lee Tuveson, *Redeemer Nation: The Idea of America's Millennial Role* (Chicago: University of Chicago Press, 1968).

3. For an attempt to deal with the balance sheet in regard to progress, see my "Progress: A Historical and Critical Perspective," in *Progress: Fact or Illusion?* ed. Leo Marx and Bruce Mazlish (Ann Arbor: University of Michigan Press, 1996).

4. See, for example, Gianna Pomata and Nancy G. Siraisi, eds., *Historia: Empiricism and Erudition in Modern Europe* (Cambridge, Mass.: Massachusetts Institute of Technology Press, 2005).

5. Quoted in *New York Times,* October 2, 2005, 3. Incidentally, Pope Benedict's remarks would seem to qualify various statements made by some of his predecessors.

6. A persuasive account can be found in Paul Gordon Lauren, *The Evolution of International Human Rights: Visions Seen,* 2nd ed. (Philadelphia: University of Pennsylvania Press, 2003).

7. For my own efforts to deal in detail with the connections between natural and human sciences, see *The Uncertain Sciences* (New Haven, Conn.: Yale University Press, 1998).

8. Eric Chaisson offers a nice treatment of some of these issues in his recent *Epic of Evolution: Seven Ages of the Cosmos* (New York: Columbia University Press, 2005).

Progress—Directionality or Betterment?

David Christian

> *Progress:* the idea "that civilization has moved, is moving, and will move in a desirable direction."
>
> *Progress:* "the idea that history is a record of improvement in the conditions of human life."[1]

The idea of *progress* contains two separable components. The first is the notion of *directionality.* The English word *progress* comes from the Latin *progressus,* a going or stepping forward.[2] The etymology implies that each forward movement depends on previous steps. In this limited sense the idea of *progress* refers to the existence of a rationally comprehensible directionality in human history.

The sense of *forward* movement helps explain the second element in the idea of *progress:* a movement *toward betterment.* This sense puts a subjective and ethical loading onto the simple idea of directionality. Early in the seventeenth century Francis Bacon had already brought these two meanings together. He insisted that human knowledge of the world had increased, and that this increase in knowledge could and should be used to improve human life. History, in short, had a direction, and that direction pointed toward betterment. Enlightenment thinkers picked up this double-barreled definition of progress with great enthusiasm, and by the nineteenth century the idea of progress held a strategic position within the human sciences.[3] Writing in the early twentieth century J. B. Bury described the "doctrine of Progress" as "the animating and controlling idea of western civilization."[4] In acquiring this ethical loading the word *progress* followed the path of other words such as *evolution* and *civilization,* both of which were pressed into service to express the Enlightenment sense of history as betterment.

From *Historically Speaking* 7 (May/June 2006)

As a first step toward clarifying the idea of progress, it is vital to separate its two component meanings. We can show many forms of directionality in human (as in biological and cosmological) history. However we know of no universally accepted criteria for *evaluating* those trends. Directionality is an objective concept that can be tested empirically. Progress is a mythic idea, one that raises ethical rather than empirical questions.

On small and medium time scales the directionality of human history is often hard to see. A historian of the Depression or the Columbian exchange or the decline of Rome will rightly reject a simple linear view of history. So will most historians operating on small scales. Besides whole areas of social life resist the idea of directionality. How could one prove that the art of the Lascaux caves is inferior to that of Picasso or that human qualities such as kindness and empathy have increased or declined over time? Nevertheless there are some clear and powerful vectors that have shaped human history and the lives of human beings in fundamental ways.

When our species first appeared between a hundred thousand and two hundred thousand years ago, it consisted of just a few tens of thousands of individuals who lived in Africa using technologies a Martian observer would have found hard to distinguish from those of other great apes. Since then humans have migrated to all parts of the Earth, and their numbers have multiplied many thousands of times to more than six billion. Meanwhile humans have greatly increased their control over planetary resources. For example energy consumption per person has multiplied from just a few thousand calories a day to more than 230,000, so that currently humans may be controlling 25 percent to 40 percent of the energy that enters the land-based biosphere from photosynthesis.[5] This represents an unprecedented increase in the ecological power and influence of a single species. Though the most spectacular changes have occurred in recent centuries, the trend is apparent throughout our history. The migrations that first took humans around the globe were largely completed before the end of the last Ice Age, and human numbers certainly increased as the range of our species widened. It was also in the Paleolithic era that humans first began to display their astonishing capacity to reshape their surroundings. Regular firing of the land transformed the environment of entire continents, particularly in newly colonized lands such as the Americas and Australia; and overhunting drove many large species to extinction in these lands.

Other long trends are equally striking. With the appearance of agriculture ten thousand years ago, humans began to transform their environments more systematically in order to maximize the production of those plant and animal species that they found most useful and eliminate those species they didn't need or value. With increasing control over natural resources, populations grew, and settlements became larger and denser. In the Paleolithic era most settlements were on the scale of an extended family. About five thousand years ago the first cities appeared, with populations of up to fifty thousand people, and today some three billion humans live in towns or cities, the largest of which contain tens of millions of people. Accelerating improvements in the technologies of communications and transportation have networked humans across the entire globe, creating intellectual synergies that were inconceivable in the Paleolithic era. Indeed much of the increase in manipulative knowledge that Bacon so admired was generated by our increasing ability to swap information, ideas, and technologies on larger and larger scales.

All in all it is foolish to deny that many fundamental features of human history are directional. Today's world is utterly different from the world of the first humans, and the differences represent something more than random fluctuations. Changes have been incremental and cumulative. Their main tendency can be described most precisely in ecological terms: One species, our own, has sharply increased its control over the resources of the planet and the biosphere, and it has done so in what is, by evolutionary standards, a very short time.

Are these trends unique to human history? In one sense they are. We know of no other large animal that has expanded its ecological power to such an extent and in such a brief era. Yet in other ways human history represents a continuation of older and larger trends. The idea that directionality in human history is part of larger trends is, of course, very ancient. It can be found within providential accounts of universal history and in the work of secular thinkers such as Herbert Spencer. In the twentieth century the human sciences largely abandoned such grand speculations. However with the consolidation of big bang cosmology in the middle of the twentieth century, it became clear that the universe itself has a history and that the history of the planet, like that of human beings, is a part of this larger story. This paradigm shift made it possible to discuss the relationship between human history and the histories of the planet and even the cosmos in scientific rather than theological or metaphysical terms.

One of the most striking findings of "big history" is that long-term trends can be identified at all these scales. For example Eric Chaisson has shown that it is possible to identify a long-term trend toward complexity.[6] The early universe was extremely simple. At the moment of the Big Bang, matter and energy could not be distinguished from each other, nor could the fundamental physical forces of gravity, electromagnetism, and the strong and weak nuclear forces. However as the universe expanded and cooled, it became more complex. Different forces assumed different roles, matter and energy went their separate ways, and temperatures began to diverge in different parts of the universe. Stars appeared, compressed and heated by the force of gravity, and as the largest stars died in supernovae, they created heavier elements, which provided the raw materials for complex chemical structures, including living organisms. The steep temperature gradients between stars and the icy cold of surrounding space also provided the energy flows needed to create and sustain complex chemicals. Indeed Chaisson has argued that in a rough and ready way, one can compare degrees of complexity by estimating the density of energy flows through different complex entities. It takes little or no energy to sustain the simple structures of interstellar space. But creating and sustaining more complex entities does require effort. If one calculates the amount of energy flowing through a given mass in a given amount of time, it turns out that energy flows are denser in living organisms than in stars and even denser in modern human society as a whole.[7] Indeed the energy flows that maintain today's global society are larger than those in any other entity that we know.

Such arguments do not imply that the appearance of our species was inevitable, of course. But they do suggest that the appearance of increasingly complex entities is part of a larger, universal trend toward denser energy flows and greater chemical complexity. On our planet essential steps along this trend include the appearance of life itself, the evolution of biological control over sunlight through photosynthesis, and the appearance of multicelled organisms, which need and can manage much-larger energy flows than single-celled organisms.[8]

Seen in the light of these larger trends, our species' astonishing control over resources appears as a striking (if localized) acceleration of an ancient and universal trend. But to understand what is unique about our species we must understand what is distinctive about this particular increase in complexity. We must identify the emergent properties that make human history different. Most historians have tiptoed around what they fear is a purely metaphysical question. However recent developments in many disciplines, from archaeology to linguistics and psychology, are helping us, through a sort of triangulation, to approach the issue as a scientific problem. And it seems that part of the answer is that we can learn collectively.[9] Equipped with a form of language that is more precise and open-ended than that of any other species, humans can share what they have learned as individuals, and they can do so with such precision that more knowledge accumulates within the collective memory than is lost. As the collective stores of knowledge have grown from the Paleolithic era to the present day, individuals have come to rely more on collective knowledge than on their individual experience in dealing with the environment and with each other. Uniquely humans face the world equipped with the experiences, the insights, and the technologies of millions of their ancestors. Over time this means that humans, as a species, have gained access to an astonishingly diverse repertoire of ideas, behaviors, and technologies. *Collective learning,* the capacity to accumulate knowledge at the level of the community or even the species, is what makes humans different. It has no real analogy among other animals. The strongest evidence for this claim is the fact that no other "cultural" species appears to have settled in environments very different from those in which it evolved. Though other species can indeed share information, they do so too inefficiently to build an accumulating store of knowledge capable of linking many individuals through time and space.

If this line of argument is accurate, it accounts for the directionality of human history. It explains how humans have accumulated the knowledge needed to exploit an increasing variety of environments with increasing efficiency, and that explains why human populations have risen and humans have gathered in denser and more complex communities. Collective learning explains why human history, unlike that of other great apes, is a history of rapid and accelerating change. It explains why there are such powerful vectors in human history—in short, why history is so clearly *directional.* Other large trends may be equally important in understanding human history. The Russian scholar V. I. Vernadsky took very seriously the idea of a universal trend toward "cephalization," or the increasing importance of brains and consciousness. This trend, he argued, would eventually lead to the construction of a "noosphere," a sphere in which consciousness would increasingly shape matter.[10]

Bruce Mazlish has referred to some extremely interesting debates within evolutionary biology about the extent and nature of directionality in evolution. Stephen Jay Gould has argued that the long-term outcomes are largely random, while Simon Conway Morris maintains that the number of available evolutionary pathways may be more strictly constrained, which suggests that if evolution occurs on many different planets, it may nevertheless follow similar pathways. Should historians be debating the directionality of human history in similar terms?[11] How can we best explain the many odd parallels between quite separate societies, such as those of Eurasia and the Americas, including the emergence of agriculture, cities, imperial states, writing, crafts, and monumental architecture? Do these suggest that the course of human history is also constrained? Is there a limited number of pathways down which collective learning could have led our species? Or would human history have turned out quite differently if rerun from the same initial starting point deep in the Paleolithic era?

The general point, I hope, is clear. Trends in human history may well be different from trends in natural history or cosmological history. But they are not *utterly* different. Above all the increasing control of energy and resources that provides the central tendency of human history represents a new (if localized) stage in trends that already existed on cosmological scales. The directionality of human history can be fully understood only in the context of these much larger trends, for the smaller arrows of human history are aligned with the planetary and cosmological vectors of big history.

Can we evaluate these trends? Can we prove that they tend towards betterment? Are the two components of progress really inseparable? Since Bacon many thinkers have assumed that increasing scientific knowledge was self-evidently good because it increased our control over the material world and promised an end to the ancient scourges of sickness, poverty, and premature death. "What gave credence to the Euro-American belief in an endlessly improving future was the rapidly accelerating expansion of knowledge of—and power over—nature achieved, during the last four centuries, by Western science, technology, economic innovation, and overseas exploration."[12]

In some areas the Enlightenment confidence in progress still seems reasonable. Our power to manipulate the natural world has increased faster than the most optimistic eighteenth-century thinkers could have imagined. As a result humans have eliminated many forms of disease and physical suffering, learned to feed much larger populations, doubled and in some regions trebled average life spans, and raised material consumption to levels that would have seemed incredible just a few generations ago. In some parts of the world societies effectively protect individuals from many of the cruder forms of oppression still prevalent in Enlightenment Europe.

However other changes have fatally undermined the easy Enlightenment confidence in science and progress.[13] Above all we have realized that there are costs to scientific progress, and they may outweigh the benefits. The wars of the twentieth century have shown that modern science can enhance our destructive as well as our creative powers. Indeed for more than half a century humans have lived with the knowledge that they have the ability at short notice to destroy themselves and everything they value, along with much of the biosphere.[14] The wars and holocausts of the twentieth century have also undermined confidence that humans will unfailingly use their increasing ecological power for betterment.

Equally unnerving is the realization that we may not be in control of the technologies we have created. Even if our intentions are good, our technologies may lead us to disaster. In a perverse twist on Smith's "hidden hand," Thomas Malthus warned early in the nineteenth century that the generous technologies that made it possible to feed more people might unwittingly create poverty on ever-larger scales. In the twentieth century environmentalism has shown that the natural environment may respond in dangerous and unexpected ways to our increasing ecological power. The scientific advances so admired in the Enlightenment may turn out to have been the result of a dangerous Faustian bargain. Do we control the large trends of human history, or do they control us? All in all the dangers of technological progress now seem at least as clear as the benefits. In such a complex situation the idea of progress *in general* has become both misleading and potentially dangerous. Instead we must analyze many different forms of directional change and assess each of them separately.

Besides what criteria could we use to evaluate progress in general? Modern historical scholarship, with its concern to understand the past on its own terms, has shown us how parochial are the standards by which we might choose to judge change. What standards can we possibly use to judge the past if, even in today's global world, there remain fundamental divisions over ethical principles? What universal ethical principles could we use to show that complex societies are objectively preferable to less-complex societies or that multicelled organisms are an improvement on bacteria? (They may be larger, perhaps, or more complex, but why should we equate either size or complexity with betterment?) As R. G. Collingwood put it: "Different ways of life are differentiated by nothing more clearly than by differences between the things that people habitually enjoy, the conditions which they find comfortable, and the achievements they regard as satisfactory. The problem of being comfortable in a medieval cottage is so different from the problem of being comfortable in a modern slum that there is no comparing them; the happiness of a peasant is not contained in the happiness of a millionaire."[15] In a famous article called "The Original Affluent Society," Marshall Sahlins argued that even by what may seem one of the clearest of criteria, that of material welfare, it is not easy to demonstrate a general trend toward betterment. For example Paleolithic diets were probably more varied and perhaps even more reliable than those of most agrarian societies, while people of the Paleolithic probably enjoyed more leisure time than middle-class citizens of today's capitalist world.[16] One of the greatest achievements of modern anthropological and historical scholarship has been to demonstrate the immense variety of cultural and ethical standards in different human societies. Yet this insight undermines the idea of progress at the deepest level, for it leaves no general criteria by which we can assess "betterment." Here is one more reason for dispensing with the idea of progress in general. History can certainly tell us about directionality, but any judgment we make about directionality depends on our own modern, culture-bound, ethical criteria.[17] Such judgments are, of course, the very stuff of modern political debate, and historians have an important role to play in such debates by clarifying the nature and power of the large trends that we can observe.

This discussion suggests that the idea of progress in general, the idea that directionality and betterment are normally aligned, is no longer tenable. In the early twenty-first century we understand better the dangers of modern technologies, and we understand how arbitrary it is to judge past societies by modern standards. On the other hand we have accumulated so much more information about the past that we can see more clearly the larger patterns and trends of human history. Understanding these trends better means that we can assess them with some precision in the light of modern ethical assumptions. But we cannot expect to re-create the broad Enlightenment consensus that history *in general* is progressive. As a coda I hope this discussion shows how the very broad perspective of big history may help us clarify the issue of directionality in human history.

Notes

1. J. B. Bury, *The Idea of Progress: An Inquiry into Its Origin and Growth* (New York: Macmillan, 1920), 2; Leo Marx and Bruce Mazlish, eds., *Progress: Fact or Illusion?* (Ann Arbor: University of Michigan Press, 1996), 1.

2. Raymond Williams, *Keywords: A Vocabulary of Culture and Society* (London: Fontana, 1976), 205–7.

3. I use the phrase *human sciences* as it is used in Bruce Mazlish, *The Uncertain Sciences* (New Haven: Yale University Press, 1998).

4. Bury, *Idea of Progress,* vii; there is a good synopsis of the history of the idea of progress in Bruce Mazlish, "Progress: A Historical and Critical Perspective," in Marx and Mazlish, *Progress,* 29–32.

5. David Christian, *Maps of Time: An Introduction to Big History* (Berkeley: University of California Press, 2004), 140–43; Vaclav Smil, *Energy in World History* (Boulder, Colo.: Westview, 1994), 236.

6. Eric Chaisson, *Cosmic Evolution: The Rise of Complexity in Nature* (Cambridge, Mass.: Harvard University Press, 2001); on big history, see Christian, *Maps of Time.*

7. Chaisson, *Cosmic Evolution,* 136–39, gives the details of these rough calculations. The energy flows (the "free energy rate density") of the sun are calculated by estimating the amount of energy emitted by the sun each second (in ergs) and dividing that figure by the sun's mass (in grams). Similarly we know how much energy it takes to keep a human being alive (about 2,800 kcal a day), and we also know the

approximate weight of an average (male) body (about 70 kg). For modern human society as a whole, Chaisson estimates the total energy consumption of humans today and divides by the average mass of human beings.

8. The story of increasing evolutionary complexity is told well in John Maynard Smith and Eörs Szathmáry, *The Origins of Life: From the Birth of Life to the Origins of Language* (Oxford: Oxford University Press, 1999).

9. I have summarized some of these arguments in *Maps of Time,* chaps. 6 and 7; see also Henry Plotkin, *The Imagined World Made Real: Towards a Natural Science of Culture* (New Brunswick, N.J.: Rutgers University Press, 2003), for a lucid and up-to-date discussion of the background to such ideas.

10. For an accessible summary of Vernadsky's ideas, see Vaclav Smil, *The Earth's Biosphere Evolution, Dynamics, and Change* (Cambridge, Mass.: Massachusetts Institute of Technology Press, 2002), 13.

11. Jared Diamond has argued forcefully that we should in his epilogue, "The Future of Human History as a Science," *Guns, Germs, and Steel* (New York: Vintage, 1998), 376–425.

12. Marx and Mazlish, *Progress,* 1.

13. The story of science's fall from grace is well told in Joyce Appleby, Lynn Hunt, and Margaret Jacob, *Telling the Truth about History* (New York: Norton, 1994), chap. 1, "The Heroic Model of Science," and chap. 5 "Discovering the Clay Feet of Science."

14. Walter Miller's 1959 novel, *A Canticle for Leibowitz* (New York: Bantam, 1997), offers a terrifying vision of an endless cycle of technological progress leading to nuclear war and historical reversal, followed by a new era of progress leading once again to nuclear war, and so on.

15. R. G. Collingwood, *The Idea of History,* ed. Jan van der Dussen, rev. ed. (Oxford: Oxford University Press, 1994), 330.

16. Marshall Sahlins, "The Original Affluent Society," in *Stone Age Economics* (London: Tavistock, 1972), 1–39. For other brief but critical discussions of long-term trends in welfare, see Stephen K. Sanderson and Arthur S. Alderson, *World Societies: The Evolution of Human Social Life* (Boston: Pearson, 2005), 36–43; and John H. Coatsworth, "Presidential Address: Welfare," *American Historical Review* 101 (1996): 1–12.

17. As Bruce Mazlish writes in the previous chapter, "Clearly it is necessary to disaggregate the notion of progress and ask in what areas, at what price, and at what stage of history."

Progress as Parochialism

J. C. D. Clark

In *Leviathan* Thomas Hobbes gave an account of his assumptions about human nature: "The Felicity of this life, consisteth not in the repose of a mind satisfied. For there is no such *Finis ultimus,* (utmost ayme,) nor *Summum Bonum,* (greatest Good,) as is spoken of in the Books of the old Morall Philosophers. Nor can a man any more live, whose Desires are at an end, than he, whose Senses and Imaginations are at a stand." Not only did desires continually extend to new things; men were also tormented about defending what they had. Power was necessary to secure what had already been won, but this, too, was involved in an endless regression. Consequently "I put for a generall inclination of all mankind, a perpetuall and restlesse desire of Power after power, that ceaseth onely in Death."[1]

With devastating candor Hobbes went at once to the heart of the problem. People are never satisfied: Once they have attained a goal, they change their goals. What they have achieved, they fear losing; what they have not achieved consumes them with desire. Within such a worldview progress, conceived as the incremental and satisfying attainment of stable and ultimate ends, was not a self-evident truth. People who think they have won the argument by saying "progress is real, but it is not inevitable" have forgotten their Hobbes. The shifting nature of ultimate ends is a familiar historical lesson. In the sixteenth century state churches were hardly reformed along Protestant lines when many of their members decided that they wished to worship in "gathered churches," rejecting the state-church ideal. In the twentieth century the Ottoman Empire was divided into secular republics; soon many of their citizens decided that they would rather live in theocratic states,

From *Historically Speaking* 7 (May/June 2006)

consciously rejecting a "modernity" now perceived as a Western imposition. Examples need not be multiplied.

The concept of *progress* was therefore constructed, like any other concept. A story of its origins was often projected back onto seventeenth- and eighteenth-century Britain, whose history was subtly misrepresented as a result. True, people then increasingly used the term *improvement,* but this was a parochial notion and was not generalized into an overall theory except for very special reasons. Yet the ideology of *progress* was eventually coined and has sometimes been internalized by the very people—the historians—who ought to stand back from it and locate it as a historical formation.

So historians who examine the idea of progress tend to fall into one of two groups. The first arranges the past as an implicit teleology; it sees humanity as engaged in a relay race, early pioneers passing the baton to later generations, who sprint across the finishing line to attain goals whose nature is held to be constant. The second group of historians dispenses with these proleptic precommitments and is therefore free to notice how late a general idea of progress was to emerge. The word *progress* was familiar from an early date; but like *race, class,* or *revolution,* it meant different things over time.

Seventeenth-century sectaries had entertained heated notions of a Second Coming, but (except in North America) these faded after 1660 and did not carry over to an idea of "progress" (even later millenarians neglected ideas of social reform; the supreme event to which they looked forward showed the unimportance of meliorist schemes). Neither Locke nor Newton predicted that the conditions of life of the mass of humanity would steadily improve. In the "battle of the books" at the end of the seventeenth century, the "moderns" claimed superiority over the "ancients" in respect of knowledge rather than gross national product. But even here it long remained possible to urge that the science of the classical world had been the origin of the achievements of the moderns, who were merely standing on the shoulders of the giants of Greece and Rome.[2]

In his great *Dictionary* (1755) Samuel Johnson offered five meanings of *progress:* four of them related to motion, one to "intellectual improvement; advancement in knowledge." None related to that "advance to better and better conditions, continuous improvement" that now features in the *Oxford English Dictionary.*[3] Similarly Johnson defined *progressive* only as "going forward; advancing"; the term carried no connotation of improvement or of a directionality to history.

The crucial stage in the development of a new meaning for the term *progress,* a general amelioration in the human condition, came when Christian aspirations for the achievement of salvation were transposed by atheists onto the everyday world. The failed Dissenting minister William Godwin, in *Political Justice* (1793), took this transposition to its logical conclusion: In the society structured according to his principles, he wrote, men "will cease to propagate" and "will perhaps be immortal"; "There will be no war, no crimes . . . no disease, no anguish, no melancholy and no resentment."[4] It was still an essentially religious vision but stripped now of its metaphysics.

When compared with the realities of poverty and desperately slow economic growth, this could only seem absurdly visionary to most people. A spiritual life beyond the material one has proved widely persuasive; since the nineteenth century the possibility of building heaven on Earth has tended to persuade only the bourgeois intelligentsia and even then chiefly as a secular distraction from the approaching fact of death. Even the lesser expectation of a general improvement in the human lot, far from being a dominant idea in eighteenth- and nineteenth-century Britain, was disputed. It functioned as a polemical assertion, not as a neutral observation of the direction taken by the economy.

British political economists in the late eighteenth and early nineteenth centuries, the period once celebrated as the scene of an "industrial revolution," were racked with doubt about the economy's future.[5] Even in the early nineteenth century, according to Boyd Hilton, most Christians (that is, most people in Britain) "saw the world as a stationary state"; for "most clerical economists . . . activity and exertion were desirable, since men should make proper use of the divine gifts of mind and muscle, but *progress,* in the sense of growth, was not the object of their economics. Movements of the economy, like those of history, were seen as cyclical rather than as linear-progressive."[6]

Even today when many more people live in relative affluence, the expansion in the world's population means that vastly larger numbers than before live in extreme poverty; whether the world average of material affluence per head has increased greatly over time is not clear. Equally unknown are subjective changes in the balance between human happiness and human misery, the "quality of life": The problem with Jeremy Bentham's "felicific calculus" is that it can never be calculated. It functions now, as it did for him, as a social polemic.

Economic and scientific development in eighteenth-century Britain (which was undoubted) did little to upset the church's teaching on the ultimately transitory nature of all human goals. Not until Darwin was the "holy alliance" of orthodox religion and natural science broken; even then Darwin could be interpreted as arguing for the endless and meaningless rise and fall of species locked in physical conflict as much as he could be held to show a general and beneficent improvement. Today we see the development of the natural sciences in general in a similar framework, not as expressing the steady linear expansion of knowledge but as the rise, dominance, and fall of successive explanatory paradigms. To use Darwinism as a simple proof of "progress" now looks historically naïve.

The most fervent believers in a generalized "progress" could only make their scenario plausible by focusing on just one time and place and implicitly making it stand for all times and places.[7] Even Edward Gibbon, who saw history as "little more than the register of the crimes, follies, and misfortunes of mankind," could not resist the temptation to depict a golden age: "If a man were called to fix a period in the history of the world, during which the condition of the human race was most happy and prosperous, he would, without hesitation, name that which elapsed from the death of Domitian to the accession of Commodus."[8] Yet for a realist like Gibbon, this happiness served only to highlight the catastrophes and sufferings of other ages, including, of course, his own.

Historical writing has developed since then; but what we cannot say is that the development of historical method reveals the steady unfolding of "progress" in the human condition. On the contrary: Microhistory is now rightly dominated by ideas of contingency; medium-level history by ideas of the counterfactual; "big history" by a growing appreciation of the cyclical nature of the natural world.

I disabuse my Kansas lecture audiences of any secular faith in progress by pointing out that if the last million years are a guide, in the next million everything they now see around them will be reduced to dust not once but eight times by the advance of glaciers more than a mile deep. At the medium level one of the greatest intellectual changes in academe in the last half century has been the undermining of the teleologies embedded in the social sciences, partly by the recovery of counterfactual analysis in history. There is no one clear direction to historical change: There are many plausible directions. Once one path of development has been taken, values are merely adapted to celebrate it.

Most historians work, of course, at the micro level. Here to acknowledge contingency is to destroy the idea of a generalized "progress": an observer who argues that progress arises through the working of contingency saws off the branch on which he or she sits. Take the case of human rights, which some commentators wish to see progressively spreading in our own day. But rights *to do* what? In a world of finite resources, one person's *freedom from* is another's *subjection to:* The rich man's freedom from constraints on burning fossil fuels is the poor man's subjection to the effects of global warming, and so on. What claims are identified as "rights" in a particular culture is a matter of contingency (some claims being preferred to others), not of eternal truth finally appreciated. Rather more sophisticated history should allow people to see how many other people fundamentally disagree with them, especially on ultimate values. That so many should so often choose to disregard this message is a remarkable and important fact.

History helps to explain why this is so. *Progress* in the modern sense was always a polemical idea. For some authors it still is, especially for militant atheists caught up in the peculiarly American debate between Darwinism and intelligent design. In a weak sense everyone now believes in progress, just as all states around the world call themselves "democracies" and claim to uphold "human rights"; this universalism means that these terms merely become synonymous with the substantive aims of those societies. A "right" is no longer right; everyone votes, but governments do as they please; peoples make progress but toward different goals. In the case of "progress," more is involved. Because "progress" was an article of a secular faith rather than an accurate comment on the nature of things, it survived such disillusioning episodes as the two world wars and the Holocaust. Like the notion of "the Enlightenment," progress was too valuable a polemical weapon to be abandoned.

Historians point out the overriding role of contingency and the counterfactual, the recurrence of war, famine, disease, and natural disaster, the inevitable end of all human endeavor with the death of the individual and the final extinction of the species; true believers in "progress" are, or profess to be, outraged. They will continue to be outraged; but this is a fact of supreme unimportance.

Notes

1. Thomas Hobbes, *Leviathan, or the Matter, Forme, & Power of a Common-Wealth Ecclesiaticall and Civil* (London: Printed for A. Crooke, 1651), part 1, chap. 11.

2. Louis Dutens, *An Inquiry into the Origin of the Discoveries attributed to the Moderns: Wherein is demonstrated, that our most celebrated Philosophers have, for the most part, taken what they advance from the Works of the Ancients* (London: Printed for W. Griffin, 1769).

3. The *Oxford English Dictionary* interestingly notes of *progress* as a verb, meaning "to carry on an action," that this usage became obsolete in England and was there readopted from America but was still "often characterized as an Americanism."

4. William Godwin, *An Enquiry Concerning Political Justice,* 2 vols. (London: Robinson, 1793), 2:871–72.

5. M. J. Daunton, *Progress and Poverty: An Economic and Social History of Britain, 1700–1850* (Oxford: Oxford University Press, 1995), 1–8.

6. Boyd Hilton, *The Age of Atonement: The Influence of Evangelicalism on Social and Economic Thought, 1785–1865* (Oxford: Clarendon, 1988), 66–67.

7. For a notable modern example of this, see especially Francis Fukuyama, *The End of History and the Last Man* (New York: Free Press, 1992).

8. Edward Gibbon, *The History of the Decline and Fall of the Roman Empire,* 6 vols. (London: Printed for Strahan and Cadell, 1776–88), 1: chap. 3.

Progress in History

Robert E. Lucas Jr.

Of course there is progress in history. Why so many people talk and write as though there were something difficult about this question is simply beyond me. Do they believe that to admit the obvious—that people as a whole are getting better off over time—will make us indifferent to the problems that still exist or blind us to environmental and other potential threats? I would imagine the opposite: The progress we see for so many millions makes it ever harder to claim that no progress is possible on the problems that remain. What we need to do about progress as historians and social scientists is not to deny its existence but describe it accurately and in detail, try to understand its character and its sources, and learn how to make the most of it.

One useful way to measure progress is to count people. Human population has been growing for as long as we have evidence, and this growth has accelerated dramatically in the postwar, postcolonial era. The nineteenth century began with fewer that one billion people and the twentieth ended with more than six billion. Since 1950 the average world population growth rate has been around 2 percent per year. This unprecedented explosion is illustrated dramatically in figure 1, taken from Robert W. Fogel's 1999 presidential address to the American Economic Association.

It is clear from figure 1 that population growth has literally turned a corner in the modern era. Is that a dark cloud on the horizon? This question merits a closer look. Figure 2, taken from my *Lectures on Economic Growth,* zooms in on the years since 1000 A.D. In my picture the population "corner" is not so sharp—I used a log scale—but it is certainly visible, sometime in the seventeenth or eighteenth century. On the same figure I also plotted total world production—real GDP. Estimating production is a more complex task

From *Historically Speaking* 7 (May/June 2006)

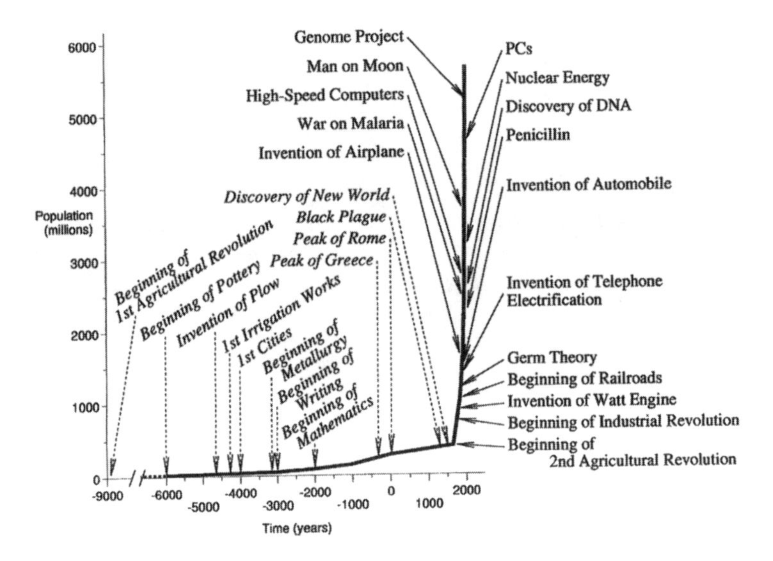

Fig. 1. The growth of the world population and some major events in the history of technology. From Robert W. Fogel, "Catching Up with the Economy," American Economic Review *89 (March 1999): 2. Courtesy of the American Economic Association*

than estimating populations, but there is no serious disagreement among specialists on the approximate accuracy of the series I have plotted. The production curve parallels the population curve over the years up to about 1800, reflecting the near constancy of living standards over these centuries (as in all earlier ones, I believe). Early in the nineteenth century the production curve turns a corner, too, and begins to grow at a much-faster rate than population does. For the first time in history the living standards of ordinary working people began to undergo sustained growth.

Far from increasing poverty the accelerating population growth of the last two centuries has been accompanied by ever-improving material conditions. In the last half of the twentieth century, growth in per capita GDP in the world as a whole reached the levels enjoyed by the leaders of the Industrial Revolution. The main feature of the postcolonial era has been and continues to be diffusion of technology to the poor economies. Over these years hundreds of millions of families have gained access, for the first time, to clean water, indoor plumbing, low rates of infant mortality, schooling, and some measure of autonomy in their life decisions. Are there value systems anywhere that would not view these developments as progress? Are there societies on record with access to these amenities that have rejected them?

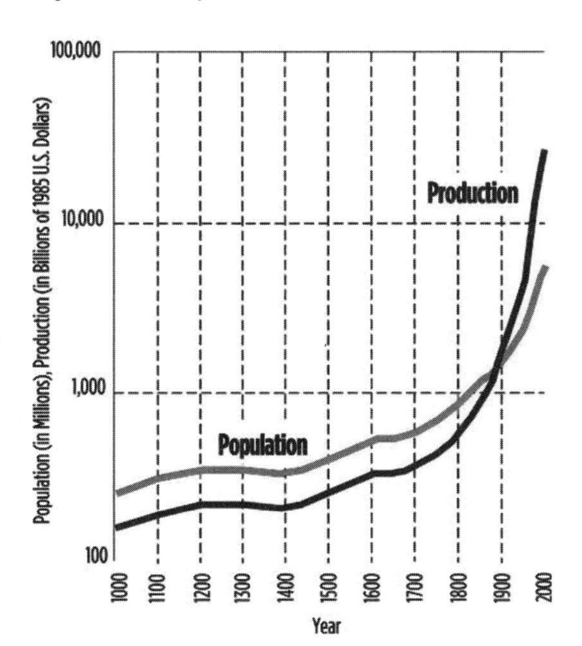

Fig. 2. World population and production. From Robert E. Lucas Jr., Lectures on Economic Growth (Cambridge, Mass.: Harvard University Press, 2002)

Many other quantitative indicators of human progress—side effects of economic growth—are available in the World Bank and various publications. Such measures can help us picture what GDP growth means, but it is also clear that not every aspect of the quality of life can be counted or measured. It is useful to have these measures, but it may be that they are best interpreted as imperfect indicators of more fundamental, underlying changes. I think this is why Fogel added markers that indicate "some important events in the history of technology" to his figure of population growth. These markers remind us that material progress measured by population and production data is primarily an intellectual achievement, a consequence of human effort and creativity.

Is there a sense in which these "events in the history of technology" themselves have what Bruce Mazlish calls "directionality"? Or is directionality only to be seen in their measurable consequences, like population and production growth? Surely we do not want to *impose* quantification on these events, as in, "On a scale of 1 to 10, how would you rate 'beginning of metallurgy' and 'discovery of DNA?'" Spurious quantification like this adds nothing to what we already know and puts a screen of unexplained opinion in between us and the facts. Nor do we *need* to assign numbers in order to define progress: The directionality in the history of ideas is much simpler and more basic. Every year new knowledge is created, and no old knowledge is lost. We have telephones and high-speed computers, and seventy years ago we only had telephones. We know the theorems of Pythagoras, Brouwer, and Banach, and Plato knew only Pythagoras's.

If this expansion in the stock of ideas is a useful way to define progress, then progress is clearly not confined to science and technology. The Greeks had Homer, and we have Homer and Shakespeare. Wagner listened to Beethoven's music, but Beethoven never heard Wagner's. Such a definition does not entail a comparison of Homer to Shakespeare or Beethoven to Wagner (whatever that might mean): Advances in the conditions of humanity do not imply or require advances in the quality of achievement of individual people. It implies only that a world with access to the latest ideas is in an important sense unambiguously better than a world with access to the earlier ones only. You would rather listen to Beethoven? Fine. But as a society we have musical opportunities that earlier generations did not have, a fact that harms no one and offers pleasure to many. If this is not progress, what is it?

I suppose this directionality in knowledge needs some qualification. Knowledge can be lost and possibly even large, valuable chunks of it. (Remember *Planet of the Apes?*) It is easier for me to think of personal examples of lost knowledge, such as the knowledge I could have obtained from conversations with my grandmother, my parents, or my youngest brother that now neither I nor anyone else will have. Some distinction between private knowledge and generally available public knowledge is needed, a distinction that would not be easy to make precise. Pursuing this distinction might clarify some of the aspects of life—loss of loved ones in my example—that are unchanging over the generations, but it does not seem likely to me that such clarification would lead to important qualifications to the directionality of the stock of available knowledge.

In the attempt to locate a more fundamental definition of progress in history, I have shifted emphasis away from quantifiable dimensions like population and GDP growth and toward a definition of progress as additions to the stock of available *ideas.* As Fogel's figure suggests, we can think of progress in the measurable indexes of success as being derived from this qualitative, intellectual progress. Doing so gets us past the idea of progress as "merely material" (or else requires us to redefine *material* to include Beethoven's music and Luther's Bible).

But a definition of *progress* confined *only* to major ideas would have limitations as serious as a definition confined only to things we can quantify. We can date when ideas first became available to someone, but we are also interested in the diffusion of ideas, their fruitful combinations with other ideas, and ultimately in their consequences for individual lives. The discovery that water can carry disease is progress, but so is the increase in the number of people who know this and who have access to clean water. It would be interesting to construct companion figures to Fogel's to illustrate the ever-widening scope of the benefits of inventive progress: plots of the number of people over time who are successfully treated with penicillin or the number of miles people have traveled by air, or the number of scholarly journal pages devoted to new mathematics. All of the good events on figure 1 initially benefited only tiny elites and then gradually over the years led to improvements in the opportunities open to hundreds of millions. Listeners of Beethoven's symphonies during his lifetime were only a small fraction of the people living in or near Vienna and a handful of other European centers. Now anyone with a CD player or iPod has essentially free access to them. An essential feature—perhaps *the* essential feature—of the Industrial Revolution has been the increasing accessibility of the fruits of technology and culture to a wider and wider set of people.

I have argued for a definition of *progress* as the expansion of the *opportunities* available to societies and to the individuals who constitute them. Defined in this way it seems to me evident that there is progress in history and that the study of its nature and sources is a worthwhile activity. It is also clear that progress is not inconsistent with the occurrence of very bad events. Expansion in opportunities admits new possibilities, but it does not rule out even the worst of the old ones.

These remarks were stimulated by Mazlish's essay, but on rereading his remarks and my own I am not sure what substantive differences I can identify or what it was that put him in such a bad temper. What is it that he values that is placed under threat by Adam Smith and Pope Benedict? I see Smith as the pioneering analyst of the progress that we can hope for in history and Pope Benedict's cited statement as a charming reminder of the unchanging aspects of the human condition.

Contingency, Necessity, Teleology, and Progress

Reply to Mazlish

Aviezer Tucker

Bruce Mazlish notes correctly that there is enormous conceptual confusion over contingency, necessity, teleology, and progress in history. I hope to clarify some of these confusions. The related concepts of contingency and necessity are independent of teleology, and teleology and progress are independent of each other as well.

Teleology and Contingency

Mazlish seems to assume that historical teleology is inconsistent with historical contingency. Because historical teleology implies that history or some historical processes have an end or a purpose, historical contingency, the possibility that things could have turned out otherwise, precludes a predestined end. Historical necessity seems to be a necessary condition for a predestined *telos,* but it ain't necessarily so.

In book 2 of his *Physics* Aristotle distinguished four types of causes: material, formal, moving, and teleological. The moving cause is the primary source of change or coming to rest, as the sculpture is to a statue. The teleological cause is the end for the sake of which an event happens, such as the finished statue for its making. In his *Timaeus* and *Philebus* Plato suggested that necessity, the moving cause (*ananke* in Greek or *causa efficiens* in Latin), may be in conflict with telos, the final cause that permeates the universe. Plato considered the moving cause to be a blind, brute, necessary force that the etiologic "Mind"—"Divine Mind," "Divine Intelligence," "Reason," "Wisdom," or plain "Zeus"—needs to "get the better of" or to "persuade" to move in the

From *Historically Speaking* 7 (May/June 2006)

correct meaningful direction, "perfection." With the exception of the Atomists, Greek philosophers felt that the necessary, moving cause was insufficient for explaining the order of the universe and the place of humanity in it. Plato and Aristotle added teleology for this purpose. Teleology is distinct from necessity, and the two may affect events in contradictory directions.

Plato's idea that teleology can use necessity as a tool was developed later in the theistic philosophies of history of Giambattista Vico and Georg Wilhelm Friedrich Hegel, where cunning divine reason uses human motivations in the service of a historical plan designed to lead to the end of history. It also appears in less metaphysically presumptuous philosophies that uphold the heterogenesis of ends—the "private vices, public virtues" social theories that Nicolas Malebranche and Vico shared with Adam Smith. Likewise Karl Marx combined necessity with teleology in arguing that inevitable economic processes would lead to a classless society, thereby eliminating the class antagonisms he saw as the necessary engine of world history. Individual voluntary action can "lengthen the birth pangs" of the end of history but not prevent it.

Conversely historical contingency can be consistent with historical teleology. If we interpret contingency probabilistically, we may suggest that at given junctures in history (social or natural) there was a certain distribution of probabilities across several options. The fact that one of these low-probability, contingent potentials actualized whereas the others did not may be consistent with either a teleological or purposeless interpretations of history. For example in a fair lottery, there is an equal distribution of very low probabilities for each ticket to be the winning ticket. It is necessary that *some* ticket should win, but it is contingent that *this* ticket will. When one ticket—say, yours—wins, it is consistent with a teleological interpretation that you were meant or destined to win.

Because both historical contingency and necessity are equally consistent with both a teleological and nonteleological interpretation of history, the concepts of teleology and contingency/necessity are independent of each other.

Teleology and Progress

Obviously not all teleological interpretations of history are progressive, because the end of history may be a return to barbarism or even the annihilation of the human race in universal entropy. Progressive interpretations of history *may be* a subset of teleological interpretations. However progressive interpretations are not necessarily teleological. It may be argued that from our value-laden perspective, without assuming that there is any grander final cause or design, history progressed for a while toward a greater fulfillment of our values. For example one may write about the history of chess or electrical engineering and notice steady progress in the quality of chess playing or engineering since the inception of these fields. Yet that does not imply that there is a telos of chess playing or engineering but merely that a mechanism of competition among chess masters and engineers, respectively, ensures that only the best survive. Over time this generates greater overall achievement. Because progressive interpretations of history are consistent with teleological and nonteleological interpretations, there is no conceptual relation between progress and teleology.

Contingency and Necessity

Yemima Ben-Menahem, following David Lewis, explicated historical contingency in terms of degrees of sensitivity to initial conditions.[1] The more sensitive a type of result is to particular initial conditions, the more contingent it is; the less sensitive, the more necessary. When several independent causal chains lead to an identical *type* of result, Ben-Menahem observes a case of historical necessity, because necessary events occur whether or not particular initial conditions are present. For example human death is necessary within certain time limits because some chain of events will bring it about irrespective of any particular initial conditions. In systems theory the most necessary overdetermined system is "hyperpredictable." In a hyperpredictable system the end state of the system, irrespective of initial conditions, is identical. Other events or processes are very sensitive to initial conditions. If these change minutely, outcomes are radically different. Events most sensitive to initial conditions are chaotic. Chaos theory describes random behavior in deterministic systems that are sensitive to minor changes in initial conditions. Deterministic laws may govern chaotic events, but their complexity usually prohibits prediction.[2] A linear, nonchaotic system may still include a critical point when one and only one parameter in that system is very sensitive to initial conditions and is governed by nonlinear algorithms.

Intuitive assessments of the degree of contingency of a system may easily lead to mistakes. The extent and location of contingency in history are theoretically contested. The Marxists and the Annales school are generally hostile to historical contingency, upholding that individuals cannot exert significant influence on the course of history. British "revisionist" historians like J. C. D. Clark think that much of history is contingent. In my opinion evaluating history's degree of contingency can only be done empirically. For instance, when historians talk of "deep" factors, they generally mean that outcomes were overdetermined, irrespective of particular "trigger" events. The conditions in the American South would have resulted in the civil rights movement after

World War II whether or not Rosa Parks sat in the front of the bus. Revisionists are right to reject teleology in historiography, such as the Whig interpretation of English history that considered constitutional democracy to be its manifest destiny. Yet rejection of destiny does not imply the endorsement of contingency. The democratization of England may have been necessary because several alternative processes were leading to it.

Historical Teleology, Weak and Strong

Historical teleology comes in a weak, value-laden, subjective version and a strong, objective version that is apocalyptic and epistemically problematic. Weak historical teleology assumes certain values, considers some historical events particularly significant, and interprets history accordingly. For example a historian may deem Wolfgang Amadeus Mozart the greatest composer of all time. He will then discuss and analyze events based on their relevance for understanding Mozart's genius. For example, the chapter on Joseph Haydn will be much larger than the one on Antonio Vivaldi. Another historian may consider Gustav Mahler to be the apex of the history. His interpretation will feature a short section on Johannes Brahms and a larger one on Alexander Zemlinsky. This type of teleology is epistemically neutral because no considerations of historical evidence, information transmission, or belief formations are relevant for judging which teleological historiography is superior to another. For example, if our Mozartian and Mahlerian historians of music meet and debate their competing interpretations, they are likely to discuss and disagree about aesthetic principles, not about historical evidence. Weak historical teleology is probably inevitable in most interpretations and narratives. As the neo-Kantians recognized more than a century ago, there must be some axiological, teleological elements in the narrative to allow for the selection and structuring of the massive amount of information we possess about the past.

Strong objective historical teleology makes the far-bolder claim that history has an exclusive objective end and that this telos can be known. Such an end may be Marx's classless society or Francis Fukuyama's liberal democracy. The problem with strong teleology is epistemic. Whether or not history has an objective end, how can we know it? If history came to an end, how would we recognize it? As Nathan Rotenstreich wrote half a century before Fukuyama, objective historical teleologies must be apocalyptic, because we can claim to understand the meaning of a process, such as a life, only when it comes to an end. The claim that history has an end can only be made at that last stage of history.[3] Accordingly all the major strong teleological interpretations of history from ancient messianism to Fukuyama have strong apocalyptic aspects. These exclusive teleological systems are inconsistent with one another. More significantly, history has had a nasty habit of continuing irrespective of the apocalyptic vision. As the late Israeli politician Josef Burg said of the Communist Party, "It will *always* be the force of tomorrow."

Strong historical teleology is baseless. Weak historical teleology, including progressive versions of it, is a reflection of the value systems of historians and as such is inevitable but does not reveal anything about history. Historical necessity and contingency, though independent conceptually of teleology, are important topics that can both reveal interesting aspects of historical processes and be ascertained empirically through the examination of historical evidence.

Notes

1. I follow in this section my extensive treatment of this topic in chapter 6 of Aviezer Tucker, *Our Knowledge of the Past: A Philosophy of Historiography* (Cambridge: Cambridge University Press, 2004).

2. Yemima Ben-Menahem, "Historical Contingency," *Ratio* 10 (1997): 99–107.

3. Nathan Rotenstreich, *Between Past and Present* (New Haven, Conn.: Yale University Press, 1958).

Rejoinder

Bruce Mazlish

Have the responses to my "Progress in History" advanced the discussion? After all the piece was intended to evoke such a possibility and was written with that aim in mind. I am pleased to say that David Christian's comment has done just that. He has either extended or clarified some of the points made in my essay (while kindly overlooking any needed corrections). He has very wisely emphasized the need to distinguish between the questions of directionality and of betterment, instancing Francis Bacon as one who in the seventeenth century sought to bring the two meanings together.

As Christian has noted, it is easier to recognize directionality on the large scale—whether in terms of cosmic evolution or more limited human evolution. With regard to the latter it is critical to recognize that there is not just one direction but many. Christian identifies a few of these over the course of the last four thousand to five thousand years. He mentions increases in human population, energy use, and the exchange of ideas. It would be hard even for those who are antagonistic to the idea of progress to argue against these facts (whether they signal change for the better is an argument separate from the directional question).

Citing Eric Chaisson Christian supports the claim that there has been a long-term trend toward complexity, that is, from simple systems to more complex ones, in all fields. Famously Stephen Jay Gould argued against this finding, but most scientists accept it. On another front complexity has become a central topic among those who discuss the ways in which the natural and human sciences can be brought closer to one another. The subject is approached by "trying to understand the dynamic behavior of complex systems that range from individual organisms to the largest economic, technical, social, and political systems."[1]

From *Historically Speaking* 7 (May/June 2006)

Here I wish to pause and consider an oft-raised question: Why has there been so much "progress" in the natural sciences and seemingly so little in the human sciences? This is a large and complex question, which I do not intend to treat here in any detail. I want only to assert that the complexity encountered in the natural sciences is simple in contrast to that characterizing human affairs. To bypass this problem attempts are frequently made to reduce the human sciences to the natural ones or subsume the former under the latter. I see little progress in this direction. As long as we remain humans, even though increasingly as prosthetic Gods, we are limited in our efforts to grapple with the complexity of human affairs.

Christian reminds us that collective learning is unique to humans. It makes cumulative knowledge possible and thus the growth in scientific findings. One form this takes is in regard to the natural sciences, another in regard to the human sciences, where because of the constant emergence of social systems and resultant change, such accumulation is far more difficult. Yet in terms of directionality it is hard to argue against the assertion that we know more in regard to "nature" than earlier generations and that, in fact, the same is true in regard to "humanity" (I put quotes around both to indicate that they are social constructions).

Does directional progress entail human betterment? No simple answer is possible. As Christian points out we witness an increase in destructive as well as constructive powers. The list is long, ranging from nuclear to climate threats and including many other dangers. Thus, as he cogently concludes, in spite of our efforts to separate descriptive from normative issues, direction from betterment, the task is more or less hopeless. He recommends, therefore, that we dispense with the idea of progress in general. Here I agree with him with one caveat. The idea of progress—some would say the myth—is a part of social reality; it serves as an inspiration for moving in a particular direction and must be studied in exactly those directional terms.

J. C. D. Clark attempts to flesh out my account of the historical origins of the idea of progress, and that is all to the good. He also reminds us that when goals are achieved, we are not satisfied but rather establish new ones. This, of course, is at the heart of modernity, but it is not clear that it characterizes nonmodern societies. Clark's point evokes thoughts about happiness. Most historians see happiness as a modern idea and one encountered mainly in the West. Psychologists and physiologists at work on the topic urge upon us the realization that much of happiness is a matter of genes and temperament once we move beyond the satisfaction of basic needs. If true (as I believe it to be), then happiness is not subject to much progress.

Favoring microhistory, in contrast to Christian's admonition that we must look mainly at large-scale events, Clark claims that small-scale history is subject to contingency (as if large-scale history were not) and then misunderstands "big history" when he reads it as embracing the cyclical nature of the natural world, thus pushing both cosmic and human evolution to the side. Reducing the idea of progress to parochialism, Clark seems to tumble into that state himself. His task, as he sees it, is to "disabuse my Kansas lecture audiences of any secular faith in progress." He makes no attempt to examine the subject in terms of directionality. Yet he mocks those "militant atheists caught up in the peculiarly American debate between Darwinism and intelligent design." Surely the militancy is mainly to be found among the proponents of intelligent design; scientists—regardless of whether they are believers, atheists, or agnostics—are concerned with the scientific understanding of evolution.

Robert Lucas, an economist, states forthrightly: "Of course there is progress in history." He thinks the question a simple one, not worthy of much thought, and the answer just as simple. Humanity's lot has been improving. I happen to believe that there is much evidence for this view, but the question is far more complicated than Lucas acknowledges. As evidence for progress he cites increased population, improved living conditions, expansion of opportunities, and advanced knowledge. Lucas also admits in passing that there are some dark sides but not to mind.

Lucas cuts away most aspects of the subject that interest historians. He is the opposite of Clark in position but not in address. I am pleased, however, that Lucas sees no "substantive differences" between us, except for my "bad temper" in regard to Adam Smith and Pope Benedict. Here he leaves me baffled. As an ardent student of Smith, I admire him too much to be uncritical.[2] Part of that esteem, however, includes the awareness that before *The Wealth of Nations* he also wrote *The Theory of the Moral Sentiments* and that the two works must be reconciled. A completely free market society was for Smith "utopian," besides being immoral. As for Pope Benedict I have difficulty in discerning bad temper in merely having quoted him.

In good temper I move on to Aviezer Tucker's "Reply." Tucker's efforts to clarify confusions over contingency, necessity, and the like do not appear to me to help solve the historian's problems. To use as an example that in a fair lottery some ticket must win, but it is contingent that this ticket will, as an analogy to the question of contingency and teleology in regard to cosmic and human evolution seems to me a way of playing philosophical games with the historian's serious questions.

In the light of the four commentaries, I wish to conclude with a few further reflections on the general topic of progress in history. The first is to consider the hoary assertion from all sides that science has nothing to do with morality. This seems to me egregiously wrong. It is the scientific method, not any particular finding, that provides us with a moral position. That method adjusted to the phenomena it is studying teaches us to be truthful about our findings, which includes a willingness to subject them to refutation. The scientific way of thinking is prepared to accept uncertainty and then by means of inquiry into the data to elaborate hypotheses, which are then further tested against the phenomena. The result is a dedication to truth where truth and a concern with reality are themselves seen as constructs and thus subject to change in the light of new evidence. This "moral" attitude is widespread in science. It is equally applicable to society, though there it encounters much-greater obstacles due to the greater complexity of the materials.

Science is based on "true witnessing," as recommended by Robert Boyle and others, but also serves as a prototype for a legal system dedicated to fair and informed verdicts.[3] It is the basis of our belief in the modern jury system in accordance with our dedication to rule by law rather than a despotic ruler. Such a method is frequently correlated with a democratic political system. Thus the same scientific method, though under greatly different conditions and convictions, permeates at least in principle both the natural and the human sciences.

Needless to say there are other claimants to moral authority. Political and religious leaders have their say as rival voices. They often claim a different kind of "truth." I want to avoid this hotbed of discord here, though I cannot refrain from referring to a quote by, in this case, a leading figure in the Muslim brotherhood, who says, "Our people look like slaves to dictatorships, and Islam came to release people from slavery to be only slaves of God."[4] In the scientific way of thinking, one is not a slave to either political or religious authority but is free to pursue the truth according to the method outlined above.

Can this be thought of as progress? A dodgy question, so let me retreat to a conclusion. History can rightly speak to the issue of directionality in human affairs and often shows movement in more than one direction. An exemplary natural scientist, Richard Lewontin, has some wise words to offer us: "We would be much more likely to reach a correct theory of cultural change if the attempt to understand the history of human institutions on the cheap, by making analogies with organic evolution were abandoned. What we need instead is the much more difficult effort to construct a theory of historical causation that flows directly from the phenomena to be explained."[5]

Notes

1. "Towards a Modern Humanism," World Knowledge Dialogue 2006, a Swiss-sponsored project to reconcile the natural and human sciences.

2. A glance at the chapter on Smith in my first book, cowritten with J. Bronowski, *The Western Intellectual Tradition: From Leonardo to Hegel* (New York: Harper, 1960), or my review essay of an edition of his *Collected Works,* cowritten with Neva R. Goodwin, an economist, for *The Harvard Business Review* (July–August 1983) will set the record straight.

3. The by-now classic account of Boyle's method is Steven Shapin and Simon Schaffer, *Leviathan and the Air-Pump: Hobbes, Boyle, and the Experimental Life* (Princeton, N.J.: Princeton University Press, 1985).

4. Quoted in the *New York Times,* March 25, 2006, 4. It should be noted that while most attention today is devoted to the return of religion (mainly, I would argue, because of its connection to politics), the long-term trend toward increased secularization is being neglected. The European Union, for example, is becoming more and more secular. Other signs of the times are the publication of books such as Sam Harris, *The End of Faith: Religion, Terror, and the Future of Reason* (New York: Norton, 2004), and Daniel C. Dennett, *Breaking the Spell: Religion as a Natural Phenomenon* (New York: Viking, 2006).

5. Richard Lewontin, "The Wars over Evolution," *New York Review of Books,* October 20, 2005, 54.

Further Readings

Engagement of Science and Religion in History

Brooke, John Hedley. *Science and Religion: Some Historical Perspectives.* Cambridge: Cambridge University Press, 1991.

Brooke, John Hedley, and Geoffrey Cantor. *Reconstructing Nature: The Engagement of Science and Religion.* Edinburgh: Clark, 1998.

Ferngren, Gary, ed. *The History of Science and Religion in the Western Tradition: An Encyclopedia.* With the assistance of Edward J. Larson, Darrel W. Amundsen, and Anne-Marie E. Nakhla. New York: Garland, 2000.

Harrison, Peter. *The Bible, Protestantism, and the Rise of Science.* Cambridge: Cambridge University Press, 1998.

———. *The Fall of Man and the Foundations of Science.* Cambridge: Cambridge University Press, 2007.

Howell, Kenneth J. *God's Two Books: Copernican Cosmology and Biblical Interpretation.* South Bend, Ind.: University of Notre Dame Press, 2002.

Larson, Edward. *Summer for the Gods: The Scopes Trial and America's Continuing Debate over Science and Religion.* Cambridge, Mass.: Harvard University Press, 1997.

Lindberg, David C., and Ronald L. Numbers, eds. *God and Nature: Historical Essays on the Encounter between Christianity and Science.* Berkeley: University of California Press, 1986.

———, eds. *When Science and Christianity Meet.* Chicago: University of Chicago Press, 2003.

Livingstone, David N. *Putting Science in Its Place: Geographies of Scientific Knowledge.* Chicago: University of Chicago Press, 2003.

Livingstone, David N., D. G. Hart, and Mark Noll, eds. *Evangelicals and Science in Historical Perspective.* New York: Oxford University Press, 1999.

Roberts, Jon H. "'The Idea That Wouldn't Die': The Warfare between Science and Christianity." *Historically Speaking* 4 (February 2003): 21–24.

Shea, William R., and Mariano Artigas. *Galileo in Rome: The Rise and Fall of a Troublesome Genius.* Oxford: Oxford University Press, 2003.

———. *Galileo Observed: Science and the Politics of Belief.* Sagamore Beach, Mass.: Science History, 2006.

Scientific Revolution

Butterfield, Herbert. *The Origins of Modern Science.* 1949. New York: Free Press, 1997.

Cohen, H. Floris. "Reconceptualizing the Scientific Revolution." *European Review* 14 (October 2007): 491–502.

———. *The Scientific Revolution: A Historiographical Inquiry.* Chicago: University of Chicago Press, 1994.

Hall, A. Ruppert. "Retrospection on the Scientific Revolution." In *Renaissance and Revolution: Humanists, Scholars, Craftsmen and Natural Philosophers in Early Modern Europe,* edited by J. V. Field and Frank James, 239–50. Cambridge: Cambridge University Press, 1993.

Harrison, Peter. "Was There a Scientific Revolution?" *European Review* 14 (October 2007): 445–57.

Heilbron, John L. "Coming to Terms with the Scientific Revolution." *European Review* 14 (October 2007): 473–89.

Lindberg, David C., and Robert Westman, eds. *Reappraisals of the Scientific Revolution.* Cambridge: Cambridge University Press, 1990.

Park, Katharine, and Lorraine Daston. "Introduction. The Age of the New." In *The Cambridge History of Science: Volume 3, Early Modern Science,* edited by Park and Daston, 1–19. Cambridge: Cambridge University Press, 2006.

Rabb, Theodore K. "The Scientific Revolution and the Problem of Periodization." *European Review* 14 (October 2007): 503–12.

Shapin, Steven. *The Scientific Revolution.* Chicago: University of Chicago Press, 1996.

Shea, William R. "The Scientific Revolution Really Occurred." *European Review* 14 (October 2007): 459–71.

Webster, Charles. *The Great Instauration: Science, Medicine, and Reform, 1626–1660.* London: Duckworth, 1975.

Yerxa, Donald A. "Historical Coherence, Complexity, and the Scientific Revolution." *European Review* 14 (October 2007): 439–44.

Progress in History

Dawson, Christopher. *Progress and Religion: An Historical Inquiry.* 1929. Washington, D.C.: Catholic University of America Press, 2001.

Fogel, Robert W. *The Fourth Great Awakening and the Future of Egalitarianism.* Chicago: University of Chicago Press, 2000.

Gray, John. *Heresies: Against Progress and Other Illusions.* London: Granta, 2004.

Harrison, Peter. "Moral Progress and Early Modern Science." *Historically Speaking* 9 (September/October 2007): 13–14.

Kuklick, Bruce. "Religion, Progress, and Professional Historians." *Historically Speaking* 9 (September/October 2007): 17–19.

Lasch, Christopher. *The True and Only Heaven: Progress and Its Critics.* New York: Norton, 1991.

Laudan, Larry. *Progress and Its Problems: Towards a Theory of Scientific Growth.* Berkeley: University of California Press, 1977.
Mazlish, Bruce, and Leo Marx, eds. *Progress: Fact or Illusion.* Ann Arbor: University of Michigan Press, 1996.
McClay, Wilfred M. "Revisiting the Idea of Progress in History." *Historically Speaking* 9 (September/October 2007): 11–12.
Nisbet, Robert. *History of the Idea of Progress.* New York: Basic Books, 1980.
Roberts, Jon H. "American Liberal Protestantism and the Concept of Progress, 1870–1930." *Historically Speaking* 9 (September/October 2007): 15–17.
Van Doren, Charles. *The Idea of Progress.* New York: Praeger, 1967.
Wright, Ronald. *A Short History of Progress.* New York: Carroll & Graff, 2005.

Contributors

John Hedley Brooke was the Andreas Idreos Professor of Science and Religion at the University of Oxford, where he was also director of the Ian Ramsey Centre and a fellow of Harris Manchester College. He is a Foundation Fellow of the Institute for Advanced Study at the University of Durham and is currently president of the International Society for Science & Religion and the U.K. Forum for Science & Religion. His main books include *Science and Religion: Some Historical Perspectives* (1991) and *Reconstructing Nature: The Engagement of Science and Religion* (1998), which was coauthored with Geoffrey Cantor, with whom he gave the Gifford Lectures at the University of Glasgow in 1995.

David Christian is professor of history at San Diego State University. He is a leading figure in the field of "Big History," which attempts to explore how human history is embedded in the histories of the biosphere and the universe. He is author of several books, including *Maps of Time: An Introduction to Big History*, winner of the World History Association's prize for the best book in world history published in 2004.

J. C. D. Clark is Hall Distinguished Professor of British History at the University of Kansas. He has made contributions to a number of fields, including British history from the earliest times to the present; religion in Britain since the Reformation; political thought; law; literature, cultural politics and the classical tradition; the "long eighteenth century," 1660 to 1832; Anglo-American and Anglo-European relations; historiography; and the history of ideas. His books include *Our Shadowed Present: Modernism, Postmodernism, and History* (2004).

Charles C. Gillispie is Dayton-Stockton Professor of History of Science at Princeton University. He is a past president of the History of Science Society, founded the Program in History of Science at Princeton University, and edited *The Dictionary of Scientific Biography*, 16 vols. (1970–80). In 1997 he became a Balzan Laureate in History and Philosophy of Science. Among his books are *The Edge of Objectivity: An Essay in the History of Scientific Ideas* (1960, 1990); *Genesis and Geology: A Study of the Relations of Scientific Thought, Natural Theology, and Social Opinion in Great Britain, 1790–1850* (1951, 1996); *Laplace, 1749–1827: A Life in Exact Science* (1997); *Science and Polity in France: The Revolutionary and Napoleonic Years* (2004); and a collection *Essays and Reviews in History and History of Science* (2007).

PETER HARRISON is Andreas Idreos Professor of Science and Religion and fellow of Harris Manchester College, University of Oxford. He has published extensively in the area of cultural and intellectual history with a focus on the philosophical, scientific and religious thought of the early modern period. His publications include *The Bible, Protestantism, and the Rise of Natural Science* (1998) and *The Fall of Man and the Foundations of Science* (2007).

EDWARD J. LARSON is University Professor and Hugh and Hazel Darling Chair in Law at Pepperdine University. He is the author of several books, including *A Magnificent Catastrophe: The Tumultuous Election of 1800* (2007), *Evolution: The Remarkable History of a Scientific Theory* (2004), *Evolution's Workshop: God and Science in the Galapagos Islands* (2001), *Trial and Error: The American Controversy over Creation and Evolution* (1985, 1989, 2003 rev. ed.) and the Pulitzer Prize–winning *Summer for the Gods: The Scopes Trial and America's Continuing Debate over Science and Religion* (1997).

DAVID C. LINDBERG is Hilldale Professor Emeritus in the History of Science at the University of Wisconsin–Madison. He is a past president of the History of Science Society and author of *The Beginnings of Western Science: The European Scientific Tradition in Philosophical, Religious, and Institutional Context* (1992) and coeditor with Ronald L. Numbers of *God and Nature* (1986) and *When Science and Christianity Meet* (2003). He is also general editor with Numbers of the eight-volume *Cambridge History of Science.*

DAVID N. LIVINGSTONE is professor of geography and intellectual history at the Queen's University, Belfast. He is vice president (research) of the Royal Geographical Society. His most recent books are *Putting Science in Its Place: Geographies of Scientific Knowledge* (2003) and *Adam's Ancestors: Race, Religion, and the Politics of Human Origins* (2008).

ROBERT E. LUCAS JR. is the John Dewey Distinguished Service Professor of Economics at the University of Chicago. He is recognized as the leader of the new classical school of economic thought and of the rational expectations theory. In 1995 he won the Nobel Prize in Economic Sciences. He is the author of *Lectures on Economic Growth* (2002).

BRUCE MAZLISH is emeritus professor of history at the Massachusetts Institute of Technology. Over a long and distinguished career he has been a pathfinder in two fields of historical inquiry: psychohistory and new global history. He has been a founding editor of two scholarly journals: *History & Theory* and *New Global Studies.* Among his most recent books is *The Idea of Humanity in a Global Era* (2008).

RONALD L. NUMBERS is Hilldale Professor of the History of Science and Medicine and of Religious Studies at the University of Wisconsin–Madison. He has written or edited more than two dozen books, including *The Creationists* (1992; expanded ed., 2006), *Darwinism Comes to America* (1998), *Science and Christianity in Pulpit and Pew* (2007), and *Galileo Goes to Jail and Other Myths about Science and Religion*

(2009). He is also general editor with David C. Lindberg of the eight-volume *Cambridge History of Science.* Numbers is a past editor of *Isis,* the flagship journal of the history of science; a past president of the American Society of Church History; a past president of the History of Science Society; and president of the International Union of History and Philosophy of Science.

William R. Shea is Galileo Professor of History of Science at the University of Padua. He is a member of the Council of the Academia Europea, past president of both the International Union for the History and Philosophy of Science and the International Academy of the History of Science, and past chair of the Standing Committee for the Humanities of the European Science Foundation. He is author, coauthor, or editor of more than twenty-five books, including (with Marianio Artigas) *Galileo in Rome: The Rise and Fall of a Troubled Genius* (2003) and *Galileo Observed: Science and the Politics of Belief* (2006).

Aviezer Tucker is a lecturer in philosophy at Queen's University, Belfast. He is the author of *Our Knowledge of the Past: A Philosophy of Historiography* (2004) and other books and articles on the philosophy of history, epistemology, and social and political philosophy. He is completing a book on posttotalitarianism and is working on another, *Origins,* dealing with the inference of common causes in biology, history, textual criticism, and comparative linguistics.

Donald A. Yerxa is codirector of the Historical Society. He has been an editor of *Historically Speaking* since 2001. He is the author or coauthor of three books, including (with Karl Giberson) *Species of Origins: America's Search for a Creation Story* (2002) and *Admirals and Empire* (1991) and editor of six other volumes in the Historians in Conversation series.

Index